Managing Mass Timber

Managing Mass Timber: A Guide to Delivering Large-Scale Mass Timber Construction Projects equips both practitioners and students with in-depth operational knowledge essential for constructing large-scale mass timber (MT) structures.

As mass timber construction continues to grow at an unprecedented pace, this textbook provides readers with the foundational knowledge required to successfully deliver MT construction projects. Grounded in widely taught theoretical frameworks from construction project management, the book focuses on the operational aspects of mass timber within the context of the five core pillars of construction management: Safety, Quality, Cost, Schedule, and Sustainability.

This text also explores the historical development of mass timber, supply chain logistics, risk management, relevant codes and standards, and other key operational aspects. It integrates core principles of project management and lean manufacturing, which are fundamental both in academic settings and professional practice. To support both students and instructors, the book includes chapter summaries, test questions, and real-world case studies.

This is the first textbook designed specifically to prepare readers in architecture, engineering and construction management (AEC), and related fields to effectively understand and manage the mass timber construction process in the field.

Managing Mass Timber

A Guide to Delivering Large-Scale Mass Timber Construction Projects

Anthony Mirando, Ph.D. and
Lameck Onsarigo, Ph.D.

Routledge
Taylor & Francis Group
NEW YORK AND LONDON

Designed cover image: Lameck Onsarigo

First published 2026
by Routledge
605 Third Avenue, New York, NY 10158

and by Routledge
4 Park Square, Milton Park, Abingdon, Oxon, OX14 4RN

Routledge is an imprint of the Taylor & Francis Group, an informa business

Library of Congress Cataloging-in-Publication Data
Names: Mirando, Anthony M. author | Onsarigo, Lameck author
Title: Managing mass timber : a guide to delivering large-scale mass timber
construction projects / Anthony Mirando, Lameck Onsarigo.
Description: First. | New York, NY : Routledge, 2026. |
Includes bibliographical references and index.
Identifiers: LCCN 2025022353 | ISBN 9781032773360 hardback |
ISBN 9781032769141 paperback | ISBN 9781003482543 ebook
Subjects: LCSH: Engineered wood construction | Lumber trade |
Building materials | Building, Wooden
Classification: LCC HD9750.5 .M55 2025
LC record available at https://lccn.loc.gov/2025022353

ISBN: 9781032773360 (hbk)
ISBN: 9781032769141 (pbk)
ISBN: 9781003482543 (ebk)

DOI: 10.1201/9781003482543

Typeset in Times
by Newgen Publishing UK

Contents

About the Authors

Anthony Mirando is an Associate Professor in the College of Architecture and Environmental Design at Kent State University, where he has been teaching, researching, and writing within the Construction Management program for over ten years. Based in Kent, Ohio, he brings more than fifteen years of experience in commercial and specialized residential construction, with over $750 million in managed work. His extensive professional practice significantly contributes to his research and writing endeavors. His research focuses on sustainable building technologies, particularly mass timber construction, third-party sustainability certifications, and various project management disciplines.

He holds a Ph.D. in Geography with a focus in construction management from Kent State University, a Master of Science in Management from Colorado State University, an executive certificate in management from University of Notre Dame, a Bachelor of Science in Construction Management from Kent State University, and LEED AP-ND sustainability credentials. In addition to his academic achievements, he has been recognized with Kent State's most prestigious teaching honor, the Outstanding Teaching Award (2024). He has also earned All-Mid-American Conference Academic First Team honors in football, was selected as a National Football Foundation Fellow, and has authored numerous peer-reviewed papers, professional presentations, and conference proceedings. He is also a co-founder of Construction Consulting Services, a platform dedicated to advancing education, research, and innovation in the construction industry.

Lameck Onsarigo is a distinguished Associate Professor of Construction Management at Kent State University, specializing in sustainable construction technologies. He holds a Ph.D. in technology management from Indiana State University, a Master of Technology Management (MTM) and a Master of Business Administration (MBA) from Bowling Green State University, and a Bachelor of Science in Construction Management from Jomo Kenyatta University of Agriculture and Technology in Kenya.

His research focuses on the development and application of innovative construction technologies that promote sustainability in the industry. Key areas of his work include trenchless technologies, mass timber construction, and the integration of green technologies into building practices. He has published extensively on topics such as the impact of trenchless technologies on construction projects, legal protections in construction contracts, the effect of prevailing wages on construction projects, strategies for enhancing labor productivity, improving quality, and the role of mass timber in sustainable building practices, among other cutting-edge topics.

In addition to his academic contributions, he is an active member of the American Society of Civil Engineers (ASCE) and the North American Society for Trenchless Technology (NASTT). He is also a co-founder of Construction Consulting Services, a platform dedicated to advancing education, research, and innovation in the construction industry.

His work has made a profound impact on the construction industry, advancing sustainable practices while fostering the integration of technological advancements and environmental responsibility. His dedication to bridging these areas positions him as a leading figure in shaping the future of sustainable construction.

Foreword

As real estate development risks intensify, the design and building industry can no longer afford to rely on outdated processes. We must embrace new technology and materials to build upon the traditional building practices and transform our operations, especially how and when we collaborate.

Mass timber, paired with innovation, presents a unique opportunity to drive this transformation. Smaller, agile teams, guided by seasoned professionals, can lower costs, improve outcomes, and address both human and natural resource challenges in measurable and impactful ways.

As students of construction, you are entering the industry at a pivotal moment—on the leading edge of the most significant shift in design and construction history. The book you're about to study couldn't be timelier for your future.

The authors of this work haven't just studied mass timber—they've lived it. Through years of meetings, interviews, and hands-on collaboration, they've tackled key questions, including the cost of mass timber. The answer? The comparative cost depends on several factors—design and strategy, the team, location, labor—but most of all, it depends on you. Your knowledge and fresh perspective will shape the future of mass timber. Its success is yours to define.

Approach this book as a builder. Picture yourself on-site, assembling a precision-crafted kit of parts in sweltering heat that tests your endurance, or in bitter cold that makes you count your fingers at the end of the day. Before you get to that point, think about how you might plan and design better drawings from lessons learned in your classrooms, life experiences, and even video games or programming skills you've acquired. I encourage you to explore reports like McKinsey's State of the Construction Industry, which highlights important industry issues such as growing labor challenges. These challenges are your greatest opportunity—not just to advance your career but to help shape the places where people live, work, and call home.

Respect will take you far—respect for those who invest their capital and vision to make projects possible, and for the men and women in the field who bring those projects to life. The people, we interact with have the potential to positively impact the trajectory of our lives. For me, that person was Tom Pope, Partner at HPA Architect and Founder of TJP Architecture. With humility, gratitude, and appreciation for those who build, you will become the bridge between the past and future—where innovation meets tradition, and where simplicity unlocks new possibilities.

This book is more than just a resource—it's a blueprint for change. It has the power to shape your future and redefine how we collaborate and build.

Finally, to the authors who chose to feature our Cleveland project as one of the case studies, we owe a debt of gratitude. Be mindful of their contributions. Read this well.

James Litwin
Vice President of Construction, Harbor Bay Ventures
Founder, Timber Places

Preface

Inspiration for this textbook stems from our firsthand experience with the construction of what is now the largest mass timber structure in the United States. This opportunity provided us with a front-row seat to the intricate process of realizing a mass timber building, serving as our entry into this remarkable and rapidly evolving field.

Mass timber construction is undergoing a period of unprecedented growth and transformation. Once considered a niche material within the construction industry, mass timber is now a sought-after solution for projects ranging from low-rise buildings to iconic high rises. This surge in popularity is driven by a growing emphasis on sustainability, advancements in engineering, and a renewed appreciation for the aesthetic and structural qualities of wood. However, as mass timber adoption expands, so too does the complexity of its construction, requiring specialized knowledge and expertise.

In today's increasingly intricate construction landscape, commercial construction managers oversee a broad spectrum of deliverables. While construction is often perceived as the physical assembly of materials, it is, in reality, a multifaceted discipline that requires collaboration across diverse teams, materials, and methods. The industry is evolving rapidly, driven by technological advancements in software, hardware, materials, and processes. To be effective, project leaders must remain well-informed of emerging trends and adapt proactively to modern construction practices. As technology and sustainability goals reshape the industry, professionals must stay ahead of these developments to lead successfully.

While innovations in building systems, methods, and materials present exciting opportunities, they also introduce new risks. Construction projects are inherently complex, and mass timber brings additional considerations. Beyond the general risks of cost, schedule, safety, quality, and design, site-specific challenges further complicate project execution, necessitating specialized mitigation strategies. As new technologies emerge, construction managers must develop a comprehensive understanding of the unique risks they present.

This textbook delves into the operational aspects of mass timber construction, offering a detailed exploration of the five foundational pillars that guide every successful project: cost, schedule, quality, safety, and sustainability. Each of these pillars plays a pivotal role in shaping the outcome of a mass timber project, influencing its financial viability, structural integrity, and long-term impact on the environment and communities it serves.

This textbook is designed to provide practitioners, students, and researchers with a practical guide to understanding the complexities of mass timber construction. By emphasizing a balanced approach that considers all five pillars, we aim to inspire the continued evolution of mass timber as a viable choice for sustainable, high-quality, and efficient construction.

As you explore the principles outlined in this text, we encourage you to think not only about their practical applications but also about the broader potential of mass timber to shape the future of the built environment. We hope this book serves as a valuable resource for understanding and advancing this innovative field.

Acknowledgments

First and foremost, we would like to thank God and our families for the guidance, support, and encouragement to work on this project.

This work would not have been possible without the invaluable contributions of many dedicated individuals and organizations. Since entering the mass timber construction and manufacturing sector in 2017, we have been met with remarkable transparency, openness, and shared passion from professionals across the industry. Their enthusiasm for mass timber construction and production is truly inspiring, strengthening our confidence in the sector's continued growth and success.

The insights, data, and support provided by those listed here were instrumental in shaping this project. Each contributor has left a meaningful impact, and their expertise has been essential to our work. Beyond this project, their efforts continue to drive innovation and progress, shaping the future of the industry as a whole.

Individuals:

James Litwin, Vice President of Construction at Harbor Bay Ventures
Reed Kelterborn, Director of Education at the Softwood Lumber Board
Dr. David Kaplan, Professor in the Department of Geography at Kent State University
Dan Whalen, Founder and Managing Principal at Places Development
Derek Ratchford, Chief Executive Officer at SmartLam North America
Dr. Amanda Colucci, Manager of Accreditation at National Architectural Accrediting Board
Dr. Simon Adamtey, Associate Professor of Constuction Management at Kent State University
Alireza Heidarian, M.S. Construction Management Student at Kent State University

Organizations:

Harbor Bay Ventures
Softwood Lumber Board
ThinkWood
WoodWorks Wood Products Council
SmartLam North America
MTC Solutions

Acronyms and Abbreviations

AI	Artificial Intelligence
AITC	American Institute of Timber Construction
ALSC	American Lumber Standard Committee
ANSI	American National Standards Institute
ASTM	American Society for Testing and Materials
BOCA	Building Officials and Code Administrators International
BOF	Best Opening Face
CEE	Cleveland-Europe Express
CLT	Cross-Laminated Timber
CMAR	Construction Manager at Risk
CPM	Critical Path Methodology
DLT	Dowel-Laminated Timber
DB	Design-Build
DBB	Design-Bid-Build
DfMA	Design for Manufacture and Assembly
EPA	Environmental Protection Agency
ESG	Environmental, Social, and Governance
EWP	Engineered Wood Products
FAO	Food and Agriculture Organization
FEM	Finite Element Modeling
FIA	Forest Inventory and Analysis
FSC	Forest Stewardship Council
GHGs	Greenhouse Gases
GiRods	Glued-in Rods
Glulam	Glue-Laminated Timber
GWP	Global Warming Potential
HVAC	HeatingVentilation and Air Conditioning
IAQ	Indoor Air Quality
IBC	International Building Code
ICBO	International Conference of Building Officials
ICC	International Code Council
IPD	Integrated Project Delivery
ISO	International Organization for Standardization
IWBI	International WELL Building Institute
JIT	Just-in-Time Delivery
LCA	Life Cycle Assessment
LCPE	Life Cycle Primary Energy
LEED	Leadership in Energy and Environmental Design
LSL	Laminated Strand Lumber
LVL	Laminated Veneer Lumber

MEPFP	Mechanical, Electrical, Plumbing, and Fire Protection
MQMP	Moisture Quality Management Plan
MT	Mass Timber
MTC	Mass Timber Construction
NCR	Non-conformance Reports
NCSEA	National Council of Structural Engineers Associations
NDAA	National Defense Authorization Act
NLGA	National Lumber Grades Authority
NLT	Nail-laminated Timber
NOAA	National Oceanic and Atmospheric Administration
O&P	Overhead and Profit
PEFC	Programme for the Endorsement of Forest Certification
PEL	Permissible Exposure Limit
PPP (three P's)	People, Planet, Profit
PSF	Pounds per Square Foot
PSL	Parallel Strand Lumber
QA	Quality Assurance
QC	Quality Control
QCP	Quality Control Plan
SBCCI	Southern Building Code Congress International
SCBF	Special Concentrically Braced Frames
SFI	Sustainable Forestry Initiative
SFM	Sustainable Forest Management
SLB	Softwood lumber board
SRCW	Special Reinforced Concrete Shear Walls
TBL	Triple Bottom Line
TCC	Timber Concrete Composite
TNC	The Nature Conservancy
USGBC	United States Green Building Council
UV rays	Ultraviolet Rays
VOC	Volatile Organic Compounds
WEF	World Economic Forum
WIN	WoodWorks Innovation Network

Glossary of Terms

American Institute of Timber Construction: the national technical trade association of the structural glued laminated timber industry. Founded in 1952.

American Lumber Standard Committee: comprises manufacturers, distributors, users, and consumers of lumber. Standing committee administering the accreditation program for the grademarking of lumber produced under the system.

American National Standards Institute: a private organization that sets guidelines for services and products across the United States. Ensures products and services are safe, reliable, and of high quality.

American Society for Testing and Materials: an organization comprising of users, producers, consumers, and general interest (academics and consultants) that develop and publish technical standards for products, materials, systems, and services.

Artificial Intelligence (AI): a technology that allows machines to learn, reason, and act in a way that usually requires human intelligence. AI is a broad field that includes various industries such as data analytics, computer science, and neuroscience.

ASTM D7247: an ASTM standard that outlines the standard test Method for Evaluating the Shear Strength of Adhesive Bonds in Laminated Wood Products at Elevated Temperatures.

Best opening face: computerized model used to simulate the process of sawing lumber from softwood logs.

Building officials and code administrators international: a group of professionals who study building codes, develop new codes, and evaluate building methods to encourage worker safety. Most municipalities adopt the codes they have developed.

Carbon sink: anything that captures and retains more carbon dioxide from the atmosphere than it emits, playing a key role in reducing greenhouse gas level. Examples include forests, oceans, and soil.

Cleveland-Europe Express: the route connects Europe to the Great Lakes Region, providing transit time advantages for cargo owners.

Construction manager at risk: a delivery method in which the construction manager is responsible for completing a project with a guaranteed maximum price.

Critical path methodology: a project management technique that identifies the sequence of tasks that are required to complete a project, with the minimum amount of time.

Cross-laminated timber: wood panels made of multiple layers of nominal lumber. Used in walls, roofs, floors, and ceilings.

Design Bid Build: a delivery method that involves separate contracts for design and construction. Considered a traditional methodology.

Design Build: a delivery method that combines the design and construction of a project into one contract.

Design for manufacture and assembly: engineering method simplifying product design with the goal of making it easier to produce and assemble.

Dowel-laminated timber: consists of stacking dimensional lumber together on its edge and fitting them together with hardwood dowels.

Engineered wood products: man-made building materials made from wood particles, veneers, or strands that are bound with adhesives. Used for beams, columns, and other structural components.

Environmental Protection Agency: An independent agency of the U.S. government taxed with environmental protection matters.

Environmental, social, and governance (ESG): criteria that measure a company's impact on the environment, its social responsibility, and how well it is governed.

Finite element modeling: a numerical technique which simulates how forces are distributed across a mass timber panel, accounting for the impact of the openings and ensuring the structural integrity is maintained. This technique is crucial for ensuring that the design complies with safety standards while accommodating necessary penetrations.

Food and Agriculture Organization (FAO): a specialized agency of the United Nations (UN) founded in 1945, whose primary goal is to combat hunger, improve nutrition, and enhance food security globally.

Forest inventory and analysis: a program in the USDA forest service research and development branch that collects, processes, analyzes, and reports on data necessary for assessing the extent and condition of forest resources throughout the United States.

Forest stewardship council: an international non-profit, multistakeholder organization created in 1993 to promote responsible management of the world's forests. Offers third-party certification for certain products.

Global warming potential (GWP): a measure of the impact of different greenhouse gases on global warming over a specified time period, typically 100 years.

Glued-in rods: metallic rods embedded into the timber using a bonding agent, such as epoxy or a high-performance adhesive.

Glulam: a construction material composed of wood pieces bonded together with adhesives.

Greenhouse gases (GHGs): are atmospheric gases that contribute to the warming of a planet's surface, by trapping heat within the atmosphere.

Heating, ventilation, and air conditioning (HVAC): refers to the systems and equipment responsible for regulating indoor comfort levels in a building. These systems control temperature, humidity, and air quality to create a safe and comfortable environment for occupants.

Indoor air quality: refers to the quality of air in any given building location. Often encompasses the amount of oxygen, moisture/humidity, and the presence of airborne contaminants.

Integrated project delivery: a construction project delivery method that aims to increase efficiency and collaboration throughout a project's construction life cycle.

International building code: refers to a model building code developed by the International Code Council. Establishes minimum requirements for the design and construction of buildings to ensure public safety, health, and welfare.

International code council: a nonprofit organization that develops and publishes comprehensive building safety and construction codes including the international building code (IBC) and the international fire code (IFC).

International organization for standardization: an independent, non-governmental, international standard development organization composed of representatives of national standard organizations.

International organization for standardization 9000: a set of ISO documents help a company ensure quality assurance in production, installation, and servicing of products.

International WELL Building Institute (IWBI): a global movement whose goal is to transform health and well-being with a people-first approach to buildings, organizations, and communities.

Just-in-time delivery: a supply chain strategy that delivers products when they are needed, reducing waste and eliminating the need for inventory.

Laminated strand lumber: a composite wood-based material used for structural framing in construction.

Laminated veneer lumber: a wood product produced by bonding thin wood veneers together with adhesives.

Leadership in energy and environmental design (LEED): provides third-party sustainability ratings of commercial buildings, homes, neighborhoods, retail, health care, schools, including every phase of the building lifecycle.

Life cycle assessment (LCA): a comprehensive methodology used to evaluate the environmental impacts of a product, service, or system throughout its entire life cycle.

Life cycle primary energy (LCPE): refers to the total amount of primary energy required throughout the entire life cycle of a product, building, or system, from raw material extraction to end-of-life disposal or recycling.

Mass timber: refers to a category of engineered wood products designed for use in constructing structural components of buildings.

Mass timber construction: use of engineered wood products to build structural elements on construction projects.

National Council of Structural Engineers Associations: dedicated to advancing the structural engineering profession through education, advocacy, career development, and community.

National Defense Authorization Act: authorizes appropriations for the acquisition or modification of various military items.

National Oceanic and Atmospheric Administration: an American scientific and regulatory agency charged with forecasting weather, monitoring oceanic and atmospheric conditions, charting the seas, conducting deep sea exploration, and managing fishing and protection of marine mammals and endangered species in the United States.

Nail-laminated timber: consists of individual wood laminations placed on their edges and mechanically fastened together using nails or screws.

Non-conformance report: a quality control document used to identify and record any deviation, failure, or non-conformance from the established standards or specifications in a project or process.

Overhead and profit: refers to the combined cost of a contractors indirect business expenses like salaries, office costs, insurance, and the desired profit margin on a site-specific project.

Parallel strand lumber: engineered wood made from wood strands that are bonded with adhesive.

Permissible exposure limit: the legal limit on the amount of chemical or physical agent that a worker can be exposed to in the workplace. Limits are set by OSHA to protect workers from adverse health effects.

Pounds per square foot (PSF): a unit of pressure used to measure the force (in pounds) exerted on a given unit area (in square feet). PSF is commonly used in construction and engineering to describe the load-bearing capacity of a structure or material, indicating how much weight a structure can safely support per unit of area.

Programme for the Endorsement of Forest Certification: an international organization which promotes sustainable forest management through independent third-party certification.

Quality assurance: a proactive, process-oriented approach focused on preventing defects before they occur.

Quality control: a reactive, product-oriented approach that focuses on identifying defects and deviations as the construction work progresses, and initiating corrective actions to address identified issues.

Quality control plan: a comprehensive document designed to guide the project team in ensuring the project is completed according to the specified requirements.

Softwood lumber board: an industry-funded initiative to promote both benefits and uses of softwood lumber products in the construction industry.

Southern Building Code Congress International: organization created to facilitate open communication between building inspectors, builders, supplier, architect, engineers, and tradespeople.

Special concentrically braced frames: a type of structural steel frame designed to resist seismic loads. Uses diagonal bracing members positioned in line with columns to ensure ductile behavior and high energy absorption during large deformations.

Special reinforced concrete shear walls: concrete walls designed to resist shear, moment, and axial forces during earthquakes.

Strip analysis: a method used to evaluate the structural impact of openings on the integrity of panels, typically in mass timber or other structural systems. This analysis helps determine how the removal of material (for MEPFP penetrations or other purposes) affects the strength and performance of the panel. Strip analysis can be performed either through manual calculations or by using advanced finite element modeling (FEM).

Sustainable forest management (SFM): a dynamic and evolving concept that aims to maintain and enhance the economic, social and environmental values of all types of forests for the benefit of present and future generations.

Sustainable forestry initiative: founded in 1994 by the American Forest and Paper Association to ensure North American forest protection.

The Nature Conservancy (TNC): a global environmental organization dedicated to creating a sustainable future where both people and nature can thrive.

Timber concrete composite: construction technique that combines timber and concrete to create a multitude of load bearing structures.

Triple bottom line (TBL): a comprehensive framework that evaluates a company's performance across three key dimensions (social, environmental, and financial) to ensure sustainable and responsible business practices.

United States Green Building Council: created in 1993 by David Gottfried to promote sustainability in building design, construction, and operations.

UV (ultraviolet) rays: a type of electromagnetic radiation emitted by the sun and other artificial sources like tanning beds.

Volatile organic compounds: a group of chemicals that readily evaporate at room temperature; some are harmful to human health.

WoodWorks Innovation Network (WIN): a voluntary database that shares important mass timber project details.

World Economic Forum (WEF): an international non-governmental organization (NGO) and think tank founded in 1971 by Professor Klaus Schwab. It is dedicated to improving the state of the world by engaging leaders from the business, political, academic, and other sectors to address global challenges and shape agendas that can drive positive change.

1 Construction Management Concepts for Mass Timber

LEARNING OBJECTIVES

After studying this chapter, the reader should be able to do the following:

1. Explain the history and development of mass timber construction
2. Understand the mass timber manufacturing process and landscape
3. Understand the differences between mass timber and the other conventional methods
4. Describe the impact of mass timber on the construction industry
5. Understand the current trends in mass timber construction and the experimental applications of mass timber.

1.1 CONSTRUCTION MANAGEMENT

This text is presented through the lens of the construction management team, highlighting the key responsibilities within the five pillars of construction management. It is essential for readers to grasp the role of the construction manager and how it aligns with various industry sectors. A construction manager is a professional tasked with overseeing a project's schedule, cost, quality, safety, scope, and function. This role involves the comprehensive coordination, planning, budgeting, and supervision of every aspect of a construction project, from inception to completion.

Construction management not only requires overseeing subcontractors performing specialized tasks, but also managing key project stakeholders, including architects, designers, material suppliers, manufacturers, and owners. The construction manager serves as the central point of control for the project and plays a crucial role in ensuring its successful completion.

DOI: 10.1201/9781003482543-1

Commercial construction projects typically involve three major stakeholders:

- **Owner**: Initiates, funds, and benefits from the completed project.
- **Architect/Engineer**: Responsible for the design and technical specifications of the project.
- **General Contractor/Construction Manager**: Manages the five pillars of construction management, ensuring that the project meets its objectives.

Over time, the role of the construction manager has expanded to encompass broader responsibilities and more complex requirements. These evolving duties demand that construction managers possess specialized qualifications, honed through a combination of education and hands-on experience, to effectively collaborate with project stakeholders and drive project success. Professional construction managers employ a wide array of tools and industry standards to meet project deliverables. However, site-specific challenges often require the adoption of innovative technologies, techniques, and processes. As the demand for new types of structures grows, construction managers must remain flexible, continuously acquiring knowledge about emerging trends and technologies to navigate these changes and ensure project success.

1.1.1 The Construction Industry

The construction industry is a vast and dynamic sector that plays a crucial role in the global economy. It is continually evolving to meet the growing demands of the built environment. In the United States alone, the industry employs 8 million people and generates nearly $2.1 trillion worth of structures annually (AGC, 2023). The outlook for the construction industry remains robust, with expected growth across various building sectors. As this growth accelerates, the demand for skilled professionals will continue to rise, further driving the need for expertise across the field.

At its core, the construction industry involves a wide array of methods, sectors, and factors that contribute to the development, maintenance, and repair of structures. Defined as the economic sector responsible for constructing, maintaining, and repairing buildings and infrastructure, the construction industry encompasses a range of internal sectors, each with distinct management requirements and operational scopes. This section explores these major sectors while examining mass timber (MT)'s role and relationship within each of them.

Currently, MT components dominate the commercial construction sector, which holds the largest market share for this material. Although various sectors within the construction industry are beginning to explore and incorporate MT, this text focuses primarily on its applications in commercial-scale projects. The major construction sectors are described in the following sections.

1.1.1.1 Commercial Sector

The commercial construction sector encompasses a wide range of building types, each often requiring a dedicated project management team to oversee the entire construction process. These projects typically involve substantial capital investment, careful planning, and the mobilization of significant resources. The construction of commercial structures often demands specialized materials, heavy equipment, and the collaboration of various professionals who provide expertise tailored to the specific needs of the project. Project initiators, or owners, are typically developers or private/public entities who drive the project forward and ensure its execution. Examples of commercial structures include:

- Office buildings
- Shopping centers
- Warehouses and distribution centers
- Hospitals
- Restaurants
- Hotels
- Multi-family housing units

When considering the scale of commercial structures, particularly in the context of MT, it's important to recognize that the material's benefits are most effectively realized in projects of the appropriate size and in certain industries. The decision to use MT often hinges on both the building type and its scale. Construction costs typically serve as a primary indicator when categorizing projects into three distinct scales, though these classifications can vary somewhat depending on the context. Nonetheless, they help define the scope of the project and the resources needed.

- **Small-scale projects** typically involve shorter timelines and require fewer resources. These may include interior and exterior renovations, updates to existing building systems, smaller additions to existing structures, or smaller free-standing structures.
- **Medium-scale projects** often entail more complex work, such as expanding or upgrading existing buildings or constructing new ones from the ground up. These projects generally require larger teams and more time to complete.
- **Large-scale projects** are typically multi-story buildings or expansive developments that require heavy equipment, specialized labor, and several years to complete. These are usually the most intricate and resource-intensive projects.

Understanding the scale of a project is crucial when evaluating the suitability of MT. The material's benefits—such as sustainability, speed of construction, and aesthetic appeal—are most pronounced in projects that are appropriately sized, making it an optimal choice for medium to large-scale commercial constructions.

1.1.1.2 Residential Sector

The residential construction sector is dedicated to the building, renovation, and repair of homes and other residential spaces. This includes a diverse range of projects, from single-family homes to apartments, town homes, and mobile homes. Larger condo buildings, multi-family structures, and residential developments can sometimes fall under the commercial construction category due to their scale and complexity.

In recent years, MT has gained popularity in multi-story, multi-family, and mixed-use residential projects, offering benefits such as sustainability, design flexibility, and reduced construction timelines. However, when considering MT for residential construction, key factors like occupancy type and project scale must be carefully evaluated.

Traditional residential construction often relies on stick-frame methods, which tend to be more cost-effective compared to MT. This cost efficiency can present a challenge when it comes to the competitiveness of MT, especially for smaller- or lower-budget residential projects. As a result, while MT holds promise for certain residential applications, its adoption is more commonly seen in larger, higher-end developments where its benefits—such as enhanced thermal performance, aesthetic appeal, and environmental impact—can justify the higher initial investment.

1.1.1.3 Industrial Sector

The industrial sector of construction primarily focuses on projects related to the production of goods, including factories, power plants, warehouses, and other specialized production facilities. While the demand for MT in this sector is increasing, it is still relatively smaller compared to its use in the commercial sector. Industrial construction projects typically require materials that can withstand heavy loads, extreme conditions, and high durability, which often leads to the preference for steel and concrete.

That said, MT is gradually gaining traction in industrial construction due to its sustainability, speed of assembly, and potential for carbon footprint reduction. As awareness grows and technologies evolve, MT may play a larger role in shaping the future of industrial facilities.

Subsequent sections will explore specific use cases and potential applications of MT in industrial construction, highlighting how it can be leveraged to meet the sector's unique demands and challenges.

1.1.1.4 Civil Sector

The civil construction sector is perhaps the least explored in terms of MT applications. This sector focuses on the design, construction, and maintenance of infrastructure related to water management, earthworks, and transportation systems. While MT has not yet seen widespread use in civil construction, ongoing research is investigating its potential, particularly for specialized civil projects.

MT offers unique advantages, including its sustainability, reduced weight, and ease of assembly, which could make it an appealing option for certain civil applications. However, the sector's reliance on materials like steel and concrete—due to their strength and durability—has traditionally limited MT's adoption.

As the MT industry continues to innovate and develop new materials, we are likely to see an expanded use of timber products in civil construction projects. Advancements in timber technology, such as enhanced durability treatments and improved structural performance, could make MT a more viable option for infrastructure projects that demand high strength and resilience.

1.1.2 Construction Materials for Structural Building Elements

The construction industry utilizes a broad range of materials, which can be classified as either structural or non-structural. Structural materials are designed to support various types of loads that a building or structure may experience. These include **dead loads** (the weight of the structure itself and any permanent equipment), **live loads** (such as occupants, mobile equipment, or traffic), and **environmental loads** (like wind, earthquakes, and rain). Structural materials are specifically engineered to withstand these forces, ensuring the safety and stability of the building. In contrast, non-structural materials serve other functions, such as aesthetics, insulation, or weatherproofing, but do not directly contribute to the load-bearing capacity of the structure.

The five most common structural materials used in construction are concrete, steel, wood, stone, and brick. For taller structures, concrete and steel have traditionally been the materials of choice due to their strength, durability, and ability to support large-scale loads. However, in recent years, MT has emerged as a viable option for tall structures. As the industry continues to explore and innovate with new materials, MT is gaining traction as a competitive and sustainable alternative to traditional building materials.

1.1.2.1 Concrete as a Building Construction Material

Globally, concrete is the most commonly used construction material, favored for applications such as driveways, pipes, foundations (including pad, mat, raft, basement, and piles), and other structural components like slabs, beams, and columns. Concrete is composed of a bonding agent (typically cement), fine aggregates (typically sand), coarse aggregates (gravel), and water. When the constituent materials are mixed and combined with water, the material can be poured into place, often compacted using vibration, and allowed to set and harden over time.

Structural concrete refers to concrete specifically designed to bear part of a structure's load. It typically incorporates reinforcement, such as steel or alternative materials, embedded within the concrete to enhance its tensile strength and load-bearing capacity. This combination of concrete and reinforcement forms what is known as reinforced concrete.

Concrete offers several compelling advantages that make it a preferred choice for structural applications:

- **Exceptional Compressive Strength**: Concrete is highly effective at resisting compression, making it ideal for supporting heavy loads.
- **Fire Resistance**: Concrete is naturally fire-resistant, providing added safety in buildings and structures.
- **Ease of Production**: Concrete is relatively simple and cost-effective to produce, with raw materials widely available.
- **Design Flexibility**: Concrete can be molded into complex shapes, making it suitable for a wide range of architectural designs and structural requirements.
- **Durability and Environmental Resistance**: Concrete is highly durable, offering long-term resistance to environmental factors such as moisture, temperature changes, and chemical exposure.

With its versatility, strength, and durability, concrete remains a fundamental material in the construction of buildings, infrastructure, and other essential structures.

1.1.2.2 Steel as a Building Construction Material

Steel is a vital structural material in construction, widely used in the fabrication of building elements such as beams, columns, bridges, and other infrastructure. It is produced by combining iron and carbon, with small amounts of additional elements to enhance its strength, durability, and versatility. Steel's exceptional tensile strength makes it especially well-suited for components subjected to pulling forces, such as framing elements.

Steel offers several key advantages that make it a preferred material in construction:

- **Fabrication Flexibility**: Steel can be fabricated into a wide variety of sizes and shapes, allowing for adaptability in design.
- **Hybrid Construction**: It can be used in combination with concrete and/or wood to create hybrid structures, harnessing the strengths of each material for optimal performance.
- **Increased Efficiency**: The prefabrication of steel components off-site can improve productivity, reduce on-site labor, and shorten project timelines.
- **Durability and Protection**: Steel can be coated and insulated to prevent rusting, enhance fire resistance, and prolong its lifespan.
- **Recyclability**: Steel can be recycled and adapted for other uses. Up to 93% of steel can be recycled to new steel components (AISC, 2025).

Overall, steel's combination of strength, flexibility, and durability makes it an essential material for modern construction, enabling the creation of both functional and aesthetically appealing structures.

1.1.2.3 Wood as a Building Construction Material

Wood, an organic material, is one of the oldest known construction materials, having been used for thousands of years. There are numerous tree species that can be utilized to produce wood, each offering distinct characteristics relative to strength, hardness, color, weight, etc. Wood has several remarkable qualities that make it not only an environmentally friendly material but also a versatile and enduring choice in modern construction. These include:

- **Natural Carbon Sink**: Wood acts as a carbon sink, capturing and storing CO_2, which helps in mitigating climate change.
- **High Strength-to-Weight Ratio**: Wood offers a remarkable balance of strength and lightness, making it ideal for a variety of structural applications.
- **Biodegradable**: Unlike many other building materials, wood is biodegradable, breaking down naturally when disposed of, which reduces its environmental impact.
- **Renewable Resource**: With proper forestry management practices—such as pre-harvest planning, sustainable harvesting, and replanting—wood can be continuously replenished, making it a sustainable building material.
- **Aesthetic Appeal**: The wide variety of wood species and its natural, warm aesthetic make wood an attractive choice for both functional and decorative purposes in construction.

A sawn solid piece of wood typically contains some naturally occurring and artificial strength-limiting defects such as knots, shakes, wood rot, splits, diagonal grain, machine bite or gouge, and checks. The defective area is where the wood is most likely to fail when subjected to loading. **Engineered wood products (EWPs)** aim to eradicate this weakness in lumber by spreading out those strength-limiting defects and ensuring that they are not concentrated in any one area, creating a weak failing point. The basic theory underlying EWPs is that wood is disassembled into small pieces then reassembled to form a product that is significantly stronger than a solid wood product of the same dimensions. EWPs also result in products with predictable strength characteristics and that can be structurally designed to bear loading. Examples of EWPs include structural building materials such as plywood, oriented strand board (OSB), glue-laminated timber (Glulam), laminated strand lumber (LSL), parallel strand lumber (PSL), laminated veneer lumber (LVL), wooden I-joists, and MT.

1.2 MASS TIMBER (MT) OVERVIEW

1.2.1 What Is MT?

MT is a unique class of EWPs that are typically characterized using large, multi-layered wood elements for constructing walls, floors, roofs, and other structural components. These products include cross-laminated timber (CLT), nail-laminated timber (NLT), glue-laminated timber (Glulam), dowel-laminated timber (DLT), and LVL. Due to their structural design, these EWPs can bear significant loads.

MT is made by bonding multiple solid wood panels together using nails, dowels, or adhesives, which provides the material with exceptional strength and stability. The ability of these EWPs to support large loads has enabled their use in the construction of larger and more complex structures, including high-rise buildings. Proponents of MT highlight several benefits, including higher productivity rates, biophilic properties, a lower carbon footprint, and natural aesthetic appeal, which will be explored throughout the text.

While MT components can be used to bear structural loads, they can also serve non-structural purposes, such as partition walls. In the context of building construction, a MT structure typically incorporates MT in its structural members (such as columns, beams, walls, floors, or roofs). However, most of these buildings are technically considered hybrid structures, as they require the integration of other materials like concrete and steel. For instance, foundations are usually made of concrete with steel reinforcements, and connections between timber members often involve steel fasteners or plates.

Examples of MT buildings with hybrid structures include the *GROW at Saint Louis Science Center* and the *Redmond Senior & Community Center* (Figure 1.1), which features a concrete foundation and MT superstructure with steel connections, and the *Salem, Oregon Public Works* (Figure 1.2), where MT is used for roofing beams, while the foundation and columns are made of structural concrete. These

FIGURE 1.1 Mass Timber building with concrete foundation and steel connections (Courtesy of Freres Engineered Wood).

FIGURE 1.2 Mass Timber used for roof structure (Courtesy of Freres Engineered Wood).

hybrid designs showcase the versatility of MT, integrating it effectively with other materials to meet the specific demands of modern construction.

While the development and growth of MT and its use has been mainly focused on vertical construction, the versatility of the material has led to its adoption in other areas that are discussed later in this chapter including blast shields, wind turbines, bridges, and industrial mats.

1.2.2 Types of MT

There are five main types of MT products that are used in the industry: cross-laminated timber (CLT), Dowel-laminated timber (DLT), glue-laminated timber (GLT), nail-laminated timber (NLT), and structural composite lumber (SCL). These five are discussed in this section.

1.2.2.1 Cross-Laminated Timber (CLT)

CLT is the most widely known MT product. It consists of layers of dimension lumber, oriented perpendicular to each other and glued together to form panels with superior strength (Figure 1.3). CLT typically consists of three, five, seven, or nine layers of dimension lumber. Because of the cross-laminations, CLT offers two-way span capabilities and is especially suitable for floors, walls (interior and exterior), elevator shafts, and roofs. Other uses of CLT that are discussed elsewhere in this book include bridge construction, dam construction, and industrial matting,

FIGURE 1.3 Cross-Laminated Timber (CLT) Panel.

which is used to build temporary access roads as part of a site access or ground protection plan.

Characteristics of CLT

- two-way span capabilities
- exceptional strength
- suitable for floors, walls (interior and exterior), elevator shafts, and roofs
- panels can be 2 ft, 4 ft, 8 ft, 10 ft, or 12 ft wide (depending on manufacturer and project needs)
- panels can range from 16 ft to 60 ft long (depending on manufacturer and project needs)
- three, five, seven, or nine layers typical (with a total depth of 4–12 inches)
- the boards can range from 5/8 inch to 2 inches thick and 2.5–9.5 inches wide
- typically made from softwood including pine, spruce, fir, and larch.

There are other non-conventional CLT panels including nail-bonded solid wood wall, also called Massiv–Holz–Mauer (MHM) and dowel-bonded CLT. MHM are massive cross-laminated panel systems made using dried softwood boards that are layered and bonded with aluminum nails. On the other hand, dowel-bonded CLT are cross-laminated panels made with layers of rough-sawn softwood boards that are bonded with hardwood dowels (Dawson et al., 2023).

1.2.2.2 Glulam or Glued-Laminated Timber (GLT)

Glulam consists of dimension lumber that is placed and bonded together using durable, moisture-resistant adhesives (Figure 1.4). This engineered wood is generally termed glulam when used for framing such as beams and columns, and GLT when used in plank applications such as decking.

FIGURE 1.4 Glued-Laminated Timber.

Characteristics of Glulam

- usually used for beams and columns and can be used for other elements such as floor beams and roof trusses
- is very versatile and can form simple straight beams, columns and panels, or complex, curved members
- exceptional strength capabilities
- panels are typically 2 ft or 4 ft wide
- beams and columns are typically 6–72 inches deep (depending on manufacturer and project needs)
- typically made from softwood including pine, spruce, fir, larch, and cedar.

FIGURE 1.5 Dowel Laminated Timber (DLT) (Courtesy of Natural Resources Canada).

1.2.2.3 Dowel-Laminated Timber (DLT)

DLT is made of softwood lumber boards that are stacked and bonded together using hardwood dowels (Figure 1.5). Dowels are friction-fit and do not involve any glue or press curing time. The friction-fit dowels hold the panels in place providing the required stability. Advantages of utilizing DLT are numerous and include aesthetic appeal, having an all-wood product (elimination of steel connections), ability to integrate mechanical, acoustic, and fire components, and providing flexibility for design teams. Integration and variability of design profiles are attractive options for development teams and the fully automated manufacturing process provides optimization of the DLT product.

Characteristics of DLT

- very versatile and can form simple straight panels, or curved members
- panels can be used for walls, floors and roofs, stairs, and elevator shafts
- like other methods, but unlike CLT, all lumber in a DLT panel is oriented in the same direction
- lumber pieces used typically range from 2 to 6 inches wide
- panels are typically 2–12 feet wide and up to 40 feet long
- typically made from softwood including spruce, pine, fir (SPF), cedar and larch, friction fit with hardwood dowels
- panels can be designed to accommodate mechanical services and fire/sound absorbing insulation (see Figure 1.6 for acoustic DLT).

1.2.2.4 Nail-Laminated Timber (NLT) or Nail-lam

NLT consists of dimension lumber that is stacked on edge and nailed/screwed together to form a larger structural member (Figure 1.7). NLT members can be

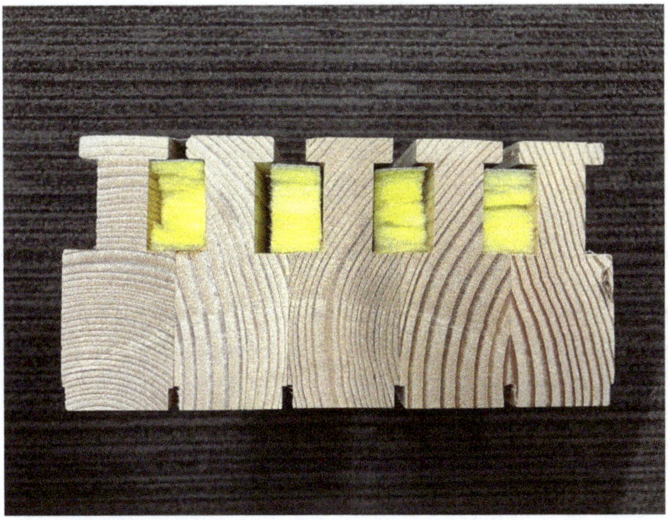

FIGURE 1.6 Acoustic DLT.

FIGURE 1.7 Nail-Laminated Timber (NLT) Panel (Courtesy of Natural Resources Canada).

pre-manufactured or easily manufactured on site because specialized equipment is not required.

Characteristics of NLT

- suitable for floors and roofs, but can also be used for walls (interior and exterior), and elevator/stair shafts
- can be prefabricated or fabricated on-site
- 2-inch, 3-inch, or 4-inch-thick dimension lumber used. Two inches thick is the most used
- 2- to 12-inch-wide dimension lumber used and fastened together with nails (no glue). 4-inch, 6-inch, and 8-inch are the most common
- lumber sizes can be altered to achieve other purposes including acoustics or to accommodate other services
- can be made using any softwood lumber.

1.2.2.5 Structural Composite Lumber (SCL)

SCL is a term used to refer to a family of EWPs that are manufactured by layering dried wood veneers, strands or flakes with moisture-resistant adhesives forming large blocks of material called billets. These billets are then cured in a heated press before being re-sawn to size.

SCL is produced at very low moisture content to limit shrinkage and its associated defects post-installation. SCL can be used for a range of structural applications including beams, joists, studs, columns, and rafters.

Characteristics of SCL

- offer predictable strength and stiffness properties and dimensional stability
- produced at very low moisture content with limited post-installation defects such as warping and splitting
- produced in a wide range of sizes (up to 24 inches deep and 65 ft long)
- can be nailed or bolted together to form built-up beams
- ensure efficient utilization of forest resources by enabling large members to be made from relatively small trees.

There are different types of SCL including:

1.2.2.5.1 Laminated Veneer Lumber (LVL)

LVL is the most widely used SCL. These are made by bonding thin wood veneers in layers to form billet, which are then cut to size (to meet the desired application). In LVL, the grain of the veneers is parallel the long side of the product (also called parallel lamination), making it capable of being loaded on its short edge when used as a beam or on its wide face when used as a plank (Figure 1.8). LVL is primarily used as structural framing for residential, commercial and industrial construction.

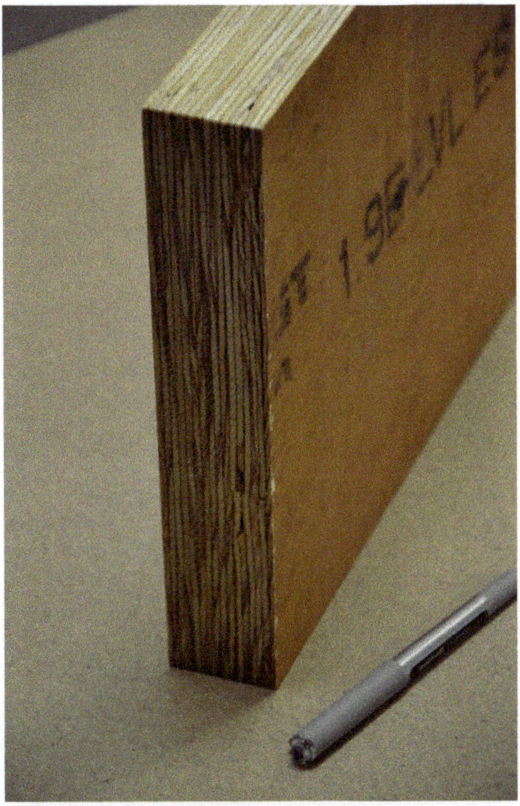

FIGURE 1.8 LVL (Courtesy of Natural Resources Canada).

1.2.2.5.2 Parallel Strand Lumber (PSL)

PSL is manufactured from wood strands laid in parallel formation and bonded together with an adhesive to form large billets that are pressed together, cured, and then cut to form the finished structural section (Figure 1.9). The length-to-thickness ratio of the strands in PSL is approximately 300 (twice as long as those to manufacture LSL and much longer than those used in the manufacture of oriented strand lumber (OSL)). PSL is primarily used as structural framing for residential, commercial and industrial construction.

1.2.2.5.3 Laminated Strand Lumber (LSL)

Its manufacturing process is similar to PSL. However, it is not made from veneers, but from flaked wood strands (like OSB) that are bonded together with an adhesive and formed into a large mat or billet and pressed (Figure 1.10). The flaked wood strands in LSL have a length-to-thickness ratio of approximately 150 and are also

FIGURE 1.9 PSL (Courtesy of Natural Resources Canada).

FIGURE 1.10 LSL (Courtesy of Natural Resources Canada).

arranged parallel to the longitudinal axis of the member. LSL is used in a variety of applications including studs, headers and beams, millwork components, window frames, and sill plates.

1.2.2.5.4 Oriented Strand Lumber (OSL)

OSL is manufactured in similar fashion to LSL, however, the flaked wood strands used in OSL have a length-to-thickness ratio of approximately 75 (shorter than those in LSL). OSL is also primarily used as structural framing for residential, commercial, and industrial construction.

1.2.2.6 Other Types of MT

1.2.2.6.1 Mass Plywood Panels (MPP)

MPP or mass ply lams (MPL) are veneer-based laminated panels. This relatively new product is an alternative to CLT and Glulam that replaces the dimension lumber used in CLT/Glulam production with veneer (see Figure 1.11). Manufacturers of this alternative have argued that this material offers some advantages over the conventional CLT/Glulam which include superior performance, slower char rate, more consistent fire performance, and better dimensional stability with lower moisture content.

FIGURE 1.11 MPP (Courtesy of Freres Engineered Wood).

1.2.2.6.2 Timber-Concrete Composite (TCC)

TCC is made from MT panels that are joined to a concrete slab by shear connectors forming a composite floor that is stronger than the non-composite alternatives. TCC panels can be made from any MT types including DLT, CLT, and glulam. TCC panels are typically used on floors or roofs but can also be used as wall panels. The timber-to-concrete shear connection can be made in a variety of ways, including the use of screws, notches, dowels, steel plates, or even alternating laminations. There are several benefits of TCCs including:

- improved sound insulation (primary consideration especially on floor applications)
- enhanced strength and durability
- enhanced fire resistance
- increased efficiency during installation
- lightweight when compared to concrete floors (about half the weight of concrete).

1.2.2.7 MT Types Summary

Like other technologies, MT is continuously evolving with progressively new types of MT and iterations of the existing types being developed such as MHM and MPP which were introduced in this chapter. The types discussed here are foundational. However, we expect more developments and iterations of these in the future. Project teams are advised to seek professional input on MT products that meet the specific project needs, including exploring the capabilities and limitations associated with each material.

1.2.3 The Historical Development of MT

Contrary to the consensus that CLT originally emanated from central Europe, the first official patent for cross-laminated panel assemblies was filed in the United States in 1920 (see Figure 1.12) and approved on August 21, 1923 (Roche, 2023). However, not much came of the approved application by Frank J. Walsh and Robert L. Watts, as production never reached impactful levels nor widespread adoption.

Meaningful development of CLT would not come again until France awarded a patent in 1985 for the production of CLT panels. This development spurred rapid growth across central Europe. 1993 saw the first residential CLT projects being in Switzerland and Germany, respectively. Early efforts were residentially focused, predating commercial scale projects. Research, development, and construction continued to proliferate across both disciplines, and countries in central Europe. Dr. Gerhard Schickhofer presented his research on the performance of CLT panels in Austria just one year later (1994), and his groundbreaking work led to approvals from multiple EU and Austrians governments to deploy CLT at the commercial

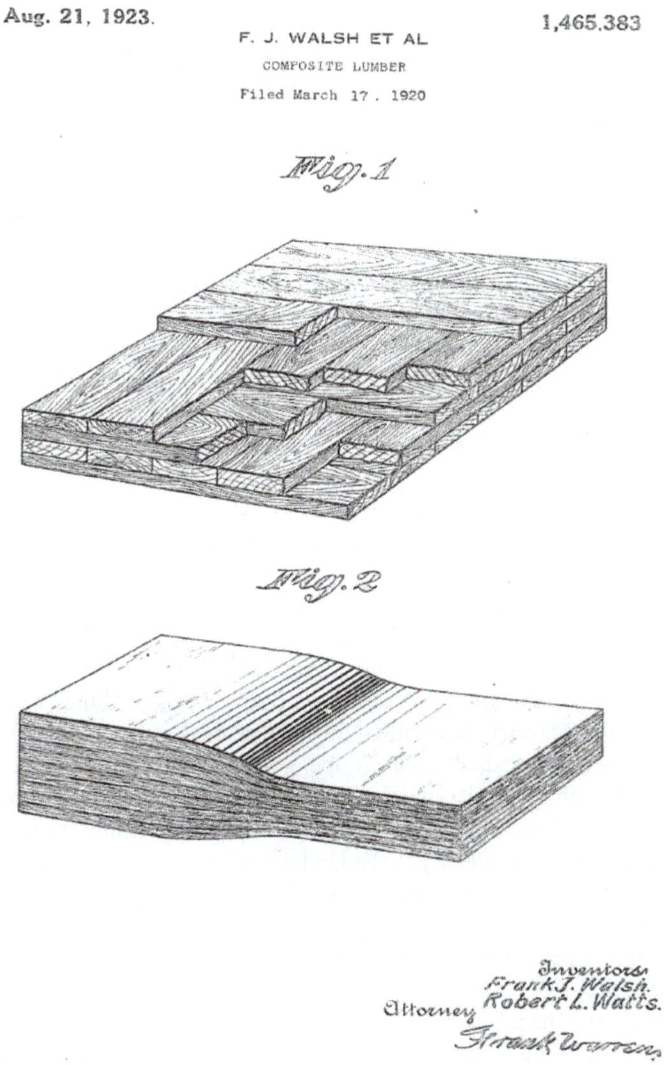

FIGURE 1.12 The First Patent Application for CLT (Courtesy of the United States Patent and Trademark Office).

level. In conjunction with improved press and manufacturing technology, the stage was set for rapid growth across the continent in the mid to late 90s.

In 1998, the first multi-story residential building was completed in Austria, with an increased level of manufacturing and raw material investment happening concurrently. As development continued, an increase in manufacturing efficiency, product approvals, improved marketing, and specific research on the sustainable attributes of MT continued to fuel expansion. The early 2000s marked significant milestones. Over 500 CLT buildings were constructed in the United Kingdom. Notably, pilot programs began in the United States and Canada in the same period. Product standards were officially approved (PRG 320) in North America (2012), with full incorporation into the Canadian National Standard for Engineering Design (2014). The same year saw the completion of the first CLT building in North America: *The Wood Innovation and Design Center* in Prince George, British Columbia, Canada.

The completion of the Wood Innovation Design Center marked the beginning of the proliferation of utilization across Canada and the United States. Expansion quickly followed in the form of official incorporation into the National Design Specification for Wood Construction in the United States (2015). Investment in research on various fronts including product development, structural design, fire testing, and seismic performance was evidenced across the United States. Other examples of widespread acceptance include the United States Department of Agriculture's Tall Wood Building Price Competitions (2015), and the International Code Council's inclusion of MT in the International Building Code (IBC).

Significant investments in entities that perpetuate marketing, research, incentive programs, and general research knowledge continued to populate the industry. WoodWorks has been the main contributor in expanding the utilization of the product, partnering with the Softwood Lumber Board, United States Forest Service, and the Forestry Innovation Investment consortium. Created in 2013, WoodWorks has established itself as the go-to resource for commercial and multi-family building design, engineering, and construction (WoodWorks, 2024). Facilitating workshops, courses, grants, and supporting conferences, highlight their commitment to the industry in the United States.

Major commercial projects across the United States began to populate city skylines around 2016. These projects mark significant strides in building height approvals, developer interest, and code adoptions by municipalities. Examples include the 512,000 square foot INTRO project located in Cleveland, OH (Figure 1.13), and the 25-story, 493,000-square foot ASCENT project located in Milwaukee, WI (Figure 1.14). These impressive projects broke both square footage and height records for MT structures in the United States cementing MT as a viable structural option.

The proliferation of MT has incentivized worldwide investment in the harvesting and manufacturing of MT. Arguably, the most expensive investment relative to the product are the manufacturing plants that produce MT products. We have seen these begin to populate in new regions. Specifically, the United States has seen cutting edge facilities begin to come online in strategic geographic regions of the pacific northwest, and southeastern areas of the county. Examples include Smartlam's $150 million dollar investment in Dothan, Alabama's manufacturing

FIGURE 1.13 INTRO Project Located in Cleveland.

FIGURE 1.14 ASCENT Project Located in Milwaukee, WI (Courtesy of United States Department of Agriculture).

facility, Mercer's facility in Spokane Valley, Washington, and Sterling's Chicago, Illinois' fully automated MT plant.

Outside of Europe and North America, other countries are making significant investment in both raw material production and commercial construction project realization. Impressive examples include China's *Jiangsu Provincial Rehabilitation Hospital*. This project highlights China's commitment to timber structures and was fostered by the relationship between Canada Wood China and the China Academy of Building Research (CABR) (Tao, 2024).

Japanese architect Kengo Kuma utilized CLT to construct a temporary park pavilion with the goal of promoting the product across the country. Designed and constructed to be mobile, the pavilion (shown in Figure 1.15) showcases the material usage and beauty to spur excitement and investment in the product. Japan has a long history of wood construction but is now making the investment in MT structures.

FIGURE 1.15 Japanese Architect Kengo Kuma's Temporary Pavilion Erected in 2020 (Courtesy of Kobayashi Kenji).

The Tanzanian government has made significant investment in CLT projects by amending laws and implementing streamlined building approval processes for MT projects like the Fumba Town Development. The Fumba Town Development is a planned community with a goal of providing housing for 20,000 people while supporting shops, medical, and educational components.

While thousands of examples can be presented here, these projects showcase the unique investments by countries across the globe and highlight not only the adaptability of MT, but its worldwide appeal. As development continues to take place projects across the world are poised to take advantage of the benefits of MT construction, perpetuating the product even further. The history of MT construction is rich and diverse, with an even brighter future.

1.2.4 Benefits and Drawbacks of MT Construction

This section provides an overview of the general benefits and drawbacks of using MT in large-scale commercial projects. While specific examples will be explored in the relevant chapters, the key points are summarized here for easy reference and dissemination. By applying the core content of this text, project teams can maximize the benefits outlined here. However, if teams do not proactively manage and address the potential drawbacks of MT, challenges are more likely to arise. Having a general understanding of both the benefits and drawbacks will equip teams to navigate the subsequent sections effectively.

1.2.4.1 Benefits

1.2.4.1.1 Sustainability

There are several misconceptions surrounding the harvesting of trees for construction materials, but extensive research underscores the significant sustainability benefits of using MT in construction.

- **Carbon Sequestration**: One of the most compelling advantages of MT is its carbon sequestration properties. Trees harvested for MT capture and store carbon throughout their growth, effectively turning buildings made from this material into "carbon banks." This contrasts sharply with traditional building materials like steel and concrete, which not only emit carbon during production but do not store it. MT's ability to lock away carbon for the life of the building makes it a powerful tool in combating climate change.
- **Biophilic Benefits**: MT also offers biophilic qualities that can enhance the health, well-being, and productivity of building occupants. Exposure to natural wood has been shown to have positive physiological and psychological effects, such as reducing stress, improving air quality, and boosting cognitive performance. These biophilic benefits contribute to a more pleasant and productive environment within MT buildings, making them particularly desirable for offices, schools, healthcare facilities, and other spaces where human well-being is a priority.

- **Sustainability Certifications**: MT buildings have a strong track record of aligning with third-party sustainability rating systems. Studies have shown that MT structures are ten times more likely to receive certifications like LEED (Leadership in Energy and Environmental Design) compared to conventional buildings (Mirando, Onsarigo, and Krsteska, 2024). These certifications are widely recognized as reliable indicators of a building's environmental performance, offering assurance that the construction meets high standards for sustainability in the built environment.
- **Renewable Resource**: Timber is a renewable resource, and when harvested responsibly, it can be continually sourced for construction without depleting forests. Sustainable forestry operations promote forest health and ensure the ongoing availability of this resource for future generations. If responsible harvesting practices are followed, timber can be replenished, unlike non-renewable materials like concrete and steel, making it a more sustainable option for construction practices.

Together, these factors highlight the sustainability benefits of MT, offering a greener alternative to traditional construction materials while contributing to the health of the planet and the people who inhabit these buildings.

1.2.4.1.2 Efficiency

The data are clear: MT structural elements can be erected significantly faster and with fewer workers compared to traditional construction methods. Unlike concrete which requires extensive curing and shoring, MT can be assembled quickly, reducing the overall construction timeline. This accelerated process leads to enhanced value and improved trade efficiency across the project.

MT's lightweight nature makes it easier to handle and transport, while its simple attachment properties streamline the assembly process. Additionally, the ability to begin building enclosure immediately after erection allows other trades, such as mechanical and electrical installations, to start earlier. This increased efficiency not only benefits the erection team but has a positive ripple effect across the entire project, reducing labor costs and minimizing delays. As a result, MT proves to be an effective and cost-efficient material that can help speed up construction without compromising quality.

1.2.4.1.3 Weight

While the weight differences between materials can vary based on a range of factors, it is generally accepted that MT components are lighter than their steel and concrete counterparts. This lightweight nature brings a range of significant benefits:

- **Reduction of Transportation Impact**: Due to its lower weight, MT reduces transportation costs and the associated carbon emissions. Lighter materials mean fewer trucks, less fuel consumption, and a smaller environmental footprint during delivery to the construction site.

- **Enhanced Seismic Performance**: MT's lighter weight and natural flexibility make it more resilient to seismic forces compared to heavier materials like concrete or steel. This characteristic can enhance the overall safety and stability of buildings, particularly in earthquake-prone areas.
- **Flexibility in Design Applications**: The relatively light MT components allow for greater design flexibility. Architects and engineers have more freedom to create innovative, dynamic structures without the constraints imposed by heavier materials, offering creative opportunities for aesthetic and functional design elements.
- **Reduced Foundation Requirements**: The reduced weight of MT leads to less stress on foundations, which can translate into cost savings and simpler foundation designs. Lighter structures require less extensive foundation work, reducing both the time and cost associated with this critical component of construction.

Overall, the lightweight properties of MT offer several advantages that make it a compelling choice for modern construction, contributing to both environmental sustainability and efficiency in the building process.

1.2.4.1.4 Aesthetic Appeal

While somewhat subjective, MT has been widely praised for its aesthetic qualities. When exposed, it brings a natural warmth and visual appeal to spaces that is hard to replicate with other materials. Its rich texture, grain patterns, and warm tones create an inviting atmosphere, making it particularly effective in enhancing the interior environment. These tactile and visual properties also make MT an ideal material for biophilic design, which seeks to strengthen the connection between people and nature. By incorporating natural elements into the built environment, MT fosters a deeper sense of well-being and a stronger bond with the surrounding environment.

Scientific studies are increasingly exploring the effects of MT on the human experience. A recent meta-analysis by Kremer et al. (2024) supports the positive biophilic response often attributed to MT, finding that exposure to wood can have beneficial psychological and physiological effects on occupants. These effects include improved mood, reduced stress, and heightened productivity, all of which contribute to a healthier indoor environment.

As research in this field continues to grow, peer-reviewed studies are quantifying the claims often made in marketing materials about the beauty and benefits of MT. This evidence-based approach is helping to solidify MT's reputation as not just a sustainable building material, but one that actively contributes to the health and well-being of building occupants.

1.2.4.1.5 Less Waste and Rework

The nature of MT—both the material itself and the process used to specify it—relies on a design methodology that minimizes the need for rework. Since each MT element is calculated and manufactured based on the final design specifications, there is little room for significant changes once construction begins. As a result, all design and layout decisions must be thoroughly finalized during the pre-construction phase.

One of the major benefits of this approach is the significant reduction in waste and rework. Each MT component is precisely engineered and accounted for, leaving minimal room for excess material. Unlike concrete, which often has a higher waste factor, MT's precision ensures that only the required materials are used. This minimizes material waste, which reduces the environmental impact and frees up space on the job site.

In addition to the environmental benefits, the emphasis on careful planning and coordination promotes detailed collaboration and communication among stakeholders throughout the project. Architects, engineers, contractors, and suppliers must work closely together to ensure that every piece is fabricated and delivered according to plan. This collaborative effort contributes to smoother project execution and a more efficient use of resources, making MT an attractive choice for projects that prioritize sustainability and streamlined workflows.

1.2.4.1.6 Realized Rental Rate Premiums

Construction costs are just one aspect of a developer's pro-forma that impacts the financial performance of a project. Development teams must also consider the income potential a MT project can generate. While rental rates are not the primary focus for the construction manager, it is crucial for teams to understand the financial benefits of choosing MT, as this can influence the pre-construction pricing process.

As discussed throughout this text, construction teams have an increased opportunity to collaborate with development teams early in the pre-construction phase when the potential use of MT is being considered. Construction teams that are knowledgeable about MT can offer significant value to developers by helping them understand how MT can positively impact project finances. One such factor is the potential for rental rate premiums, which can help offset initial material costs.

Recent data show that MT structures can command rental premiums ranging from 2% to 18%, depending on the geographic region (Neutral, 2024). The reasons behind these premiums are subjective but include:

1. **Visual Appeal**: The natural aesthetic of exposed timber is a major draw for renters, creating a unique and attractive environment.
2. **Sustainability**: Renters are increasingly focused on the sustainability of the spaces they occupy, and MT's eco-friendly properties make it an appealing choice.

3. **Biophilic Benefits**: The biophilic properties of MT—such as improved air quality, stress reduction, and enhanced productivity—contribute to the overall well-being of occupants and add value to the space.

By understanding these factors, construction teams can help development teams better navigate the financial implications of using MT, positioning the material as a long-term investment that can deliver strong returns through rental rate premiums.

1.2.4.2 Drawbacks to MT

This section summarizes some of the drawbacks identified when using MT in large-scale projects. While there are challenges associated with MT, part of the focus of this text is to not only present each drawback but also equip project teams with the knowledge and strategies to mitigate these issues. Though these challenges may seem significant, proactive management and planning can minimize or eliminate their impact. Additionally, many of these drawbacks are likely to diminish as project teams, insurers, municipalities, and owners become more familiar and comfortable with using MT.

It's important to note that many of the issues discussed here are primarily due to the relative infancy of MT in the construction industry. As the material gains wider adoption and the industry continues to innovate and refine processes, these challenges will become more manageable. With ongoing research, technological advancements, and improved standards, MT's potential will be better realized, and the drawbacks associated with its use will be mitigated. In the meantime, proactive planning, continuous education, and collaboration among stakeholders will help ensure that MT projects are successful and efficient.

1.2.4.2.1 Moisture Mitigation

Every commercial MT project requires a comprehensive water mitigation plan to address the effects of moisture exposure on MT elements. Wood's properties, including strength and stability, are directly influenced by its moisture content, making it essential for project teams to work closely with manufacturers and subcontractors to implement protective measures that minimize excessive moisture exposure throughout the construction process.

Effective management and mitigation strategies, which are covered in detail in the quality chapter, help ensure that MT components maintain their integrity. While these techniques may require careful planning and come with additional costs, they can be manageable when approached with foresight and proper coordination. These strategies typically include protective coverings during transportation and storage, as well as moisture monitoring throughout the construction process. By accounting for these factors early in the planning stages, teams can prevent potential issues before they arise.

In contrast, traditional structural materials like steel and concrete are generally less sensitive to moisture exposure, which makes them easier to manage in environments with fluctuating humidity or moisture levels.

1.2.4.2.2 Negotiating Local Municipalities

As a relatively new material, the use of engineered wood in major capital projects can sometimes cause hesitation among local municipality building departments. Due to its innovative nature, MT may not always fit neatly into existing building codes or regulatory frameworks, which can create uncertainty or delays in the approval process. As a result, development and construction teams may face significant upfront pre-construction costs and potential schedule delays, depending on the location and familiarity of the local municipality with MT.

In many cases, variances or exceptions to building codes may be required to meet project goals, as MT components might not always align with traditional construction standards. Regulatory uncertainty remains a challenge, especially in regions where MT is less common.

As MT becomes more widely accepted and integrated into building codes, the approval process will likely become more streamlined, reducing the burden of regulatory hurdles on future projects. In the meantime, proactive collaboration between project teams, local building departments, and regulatory bodies will be key to navigating these challenges and ensuring the successful delivery of MT projects.

1.2.4.2.3 Increased Soft Costs

General condition premiums are a significant concern for construction and development teams working with MT. While the estimating section provides a more detailed breakdown of these costs, this section offers a summary of the key factors contributing to increased expenses.

When incorporating MT into a project, it's essential to account for increased insurance premiums, pre-construction costs, and project contingencies in the cost projections. Insurers, typically risk-averse when it comes to new construction methods, often raise insurance costs due to the perceived risks associated with MT. This can drive up premiums both during construction and for post-construction coverage.

Additionally, to address the uncertainty faced by teams with limited experience using MT, construction contingencies are frequently increased. These contingencies help offset the perceived risks and potential unforeseen issues associated with working with new materials. While this may result in higher upfront costs, these premiums and contingencies are often seen as a necessary measure to manage risk until MT becomes more widely adopted and recognized by insurers and other stakeholders.

As more projects using MT are successfully completed, and the material gains broader acceptance within the industry, it is likely that insurance premiums and contingencies will decrease over time, as the perceived risks associated with the material diminish.

1.2.4.2.4 Upfront Premiums on Material

Many development teams experience "sticker shock" when receiving quotes from MT manufacturers for production and shipping. Currently, MT components used for structural applications carry about a 10% cost premium compared to traditional structural materials like steel and concrete. This premium is largely attributed to the developing domestic supply chain, which has yet to fully streamline operations and reduce production costs. Additionally, transportation costs play a significant role in the higher price, as much of the MT used on domestic projects is sourced from European suppliers, adding shipping fees and import taxes to the overall expense.

These upfront costs can be a major deterrent for development teams, with numerous projects being delayed or even halted due to the initial expense of using MT as a structural option. However, it's important to note that, like the other challenges discussed in this section, it is anticipated that these premiums will decrease over time. As the domestic supply chain matures, production processes become more efficient, and local manufacturers increase their capacity, the cost of MT is expected to become more competitive with traditional materials. Additionally, as MT adoption grows, economies of scale will likely reduce shipping costs and enhance the affordability of future projects.

1.2.4.2.5 Limited Experience

A lack of experience with MT remains a significant drawback. During the pre-construction phase, the process can become cumbersome, inefficient, and more costly as teams work to identify the best delivery methods, coordinate the necessary contractors, and ensure that all elements of the project align. The lack of familiarity with the material can result in delays and missteps in the planning phase, adding both time and financial resources to the project.

Additionally, specialty contractors involved in the installation of the MT structure, as well as other trades interacting with the material, may raise their bid prices and extend their schedules to account for their limited experience working with MT. The perceived risk and unfamiliarity with the material can lead these contractors to factor in additional time for training, adjustments, or potential complications on-site, further driving up costs and extending the timeline.

For project teams, this inexperience can present considerable challenges. It often requires more patience, detailed planning, and flexibility, as well as a larger financial cushion to address unexpected hurdles. However, these challenges can be mitigated by fostering strong collaboration between all stakeholders, including contractors, manufacturers, and designers, and by investing in training and knowledge-sharing initiatives. As experience with MT increases within the industry, these growing pains are expected to diminish, but in the short term, they remain a hurdle that teams must navigate.

1.3 MT MANUFACTURING

At its core, MT manufacturing involves the process of taking smaller wood elements and connecting them using adhesives, dowels, nails, or screws to create larger structural components. The result is a highly durable and engineered material that meets the design parameters for a specific project.

MT is produced in specialized manufacturing plants that are equipped with advanced machinery to handle various processes, including re-manufacturing, finger-jointing, and surfacing lumber. The plant also uses glue applicators, panel presses, and computer-controlled (CNC) routers, which trim panels to size and cut necessary openings for doors, windows, and other features. The production process follows a series of carefully sequenced steps, as outlined by Seagate (2018):

1. **Selecting the Lumber**: Common wood types for MT include Yellow Pine, Western Larch, Douglas Fir, and SPF Lumber, chosen for their strength, stability, and availability.
2. **Drying the Lumber**: The lumber is dried in a kiln to achieve a moisture content of approximately 12%, ensuring it is stable and suitable for use in structural applications.
3. **Removing Defects**: Defects, such as knots or cracks, are removed to ensure the integrity and strength of the final product.
4. **Trimming and Jointing**: The lumber is then trimmed and jointed to the correct lengths as required for the project.
5. **Assembling the Panels**: Boards are placed side by side to form solid wood layers that will make up the final panels.
6. **Applying Adhesive**: Adhesive is sprayed on each layer to bond the wood together securely.
7. **Pressing the Panel**: The panels are pressed together using a hydraulic press, ensuring that the layers are tightly bound and structurally sound.
8. **Fabricating the Panels**: Finally, the panels are fabricated by cutting them to size, milling them, and finishing them according to specific project requirements.

This process results in strong, reliable MT components that can be used in a variety of structural applications.

1.3.1 MT Manufacturing Technologies

Like with most facets of construction management, the manufacturing process is not immune to the injection of advanced technology to meet the need of modern-day production. Advancements in manufacturing technology have been a direct driver of MT production. The integration of advanced technology, including AI optimized manufacturing systems, has impacted fabrication in several ways, including, but not limited to increased production capacity, increased flexibility in production, higher levels of quality control, safer operations, and increased efficiency through manpower and waste reductions.

Manufacturers are typically transparent regarding the capability of their facilities and often extend invitations to project teams to evaluate production rates and capabilities. In terms of level of technological innovation, spectrums exist depending on the manufacturer. More advanced manufacturers run operations with more automation and less human interface, whereas older plants rely more on physical labor to interact with the production process. It is recommended that project teams familiarize themselves with the manufacturing process from their respective supplier. While teams do not need expertise in the process, being comfortable with the supplier through site visits can ensure schedule and quality requirements are met.

1.3.2 Global Overview of MT Manufacturing

An important component of MT construction is, not only an understanding of the production and manufacturing system, but also the logistical aspects. Integral to project success is supply chain management which is critical in ensuring on time delivery, product availability, and understanding transportation cost associated with procurement. The oldest and arguably the most advanced manufacturing facilities are located across Europe. By sheer market volume the DACH region + Italy and Czech Republic, comprising Germany, Austria, and Switzerland account for roughly 2 million m^3.

The complexity and technology integrated in these facilities is groundbreaking. Typically partnering with European manufacturers, global facilities are leveraging technology to drive efficiency in MT production. High flexibility of production, volume of production, production rates, quality, and reduced need for human workers are the touted benefits of these automated manufacturing facilities. The manufacturing facilities are in proximity to wood resource availability, drawing strategic investment to geographies with favorable wood harvesting guidelines and availability.

Longstanding experience with timber harvesting and production has placed European producers with a competitive advantage in cost, quality, design, and supply chain management. Other regions are slowly catching up to these impressive capabilities; however, it requires major monetary investment and takes time to establish industry presence. Europe's dominance is evidenced by its use in the majority of United States MT projects. As of 2024, European sourced products still retain a competitive cost advantage to United States domestic and, even longer established, Canadian manufacturers. However, as outlined in the subsequent section, U.S.-based manufacturing is catching up. It is anticipated that manufacturing and transportation costs will become more competitive with established suppliers.

While EU countries hold significant market share in production, other regions and countries are increasingly producing significant amounts of MT products.

Canadian investment in MT production and design is strong. Considered one of the major producers and users of MT, Canada has completed over seven-hundred MT projects, with more than one-hundred and forty projects either planned or under construction as of 2024. Canadian companies Kalesnikoff, Nordic Structures, and Western Archrib have invested heavily in the conversion of standard mills to now produce MT components. Canada is responsible for serving 25% of the world's MT market and is quantified by producing approximately 1 million cubic meters of product. To meet global demand, production increase of 15% is projected for 2025 (Allan & Eaton, 2024).

Production facilities are coming online yearly in geographies outside of Europe and North America. Examples include manufacturing facilities in New Zealand, Australia, Japan, and China to meet regional and global demand of the product.

Japanese company Cypress Sundaya Co. Ltd recently completed a fully automated manufacturing line that is able to use different adhesive types, finger jointing, buffering systems, edge glue presses, and other unique MT applications to offer turn key solutions for building projects.

1.3.3 Domestic Overview of MT Manufacturing

Like the countries mentioned above, the United States is investing heavily in the production of MT to compete for market share. The gap in technology is so broad that international suppliers maintain a cost advantage over current domestic suppliers, even when shipping products across oceans. However, it is anticipated that these domestic premiums will erode as the technology is refined on par with more established suppliers.

Significant investment in United States manufacturing has been made by several key companies. These include SmartLam, Sterling, Mercer, Seagate, Swinerton, and Timberlab. Internationally based firms have also entered the domestic market. Binderholz (Austria) and Structurecraft (Canada) have opened U.S.-based facilities in the Southeast region.

The map in Figure 1.16 indicates the geographical location of the top six domestic producers of MT. As stated above, location of manufacturing facilities is directly tied to raw material acquisitions. Positioning these facilities near major forested regions that can adequately support production of the material. Domestic manufacturing is having a big impact on the transportation cost associated with the material, in conjunction with eliminating costly and timely tariffs, international inspections, and custom paperwork requirements. Benefits for US-based commercial projects are numerous considering the logistical concerns associated with the importation of international materials.

Significant monetary investment in domestic manufacturing is proof that the industry believes in the benefits and demand for the product across the United States and is a key indicator as to the staying power of MT as a building material.

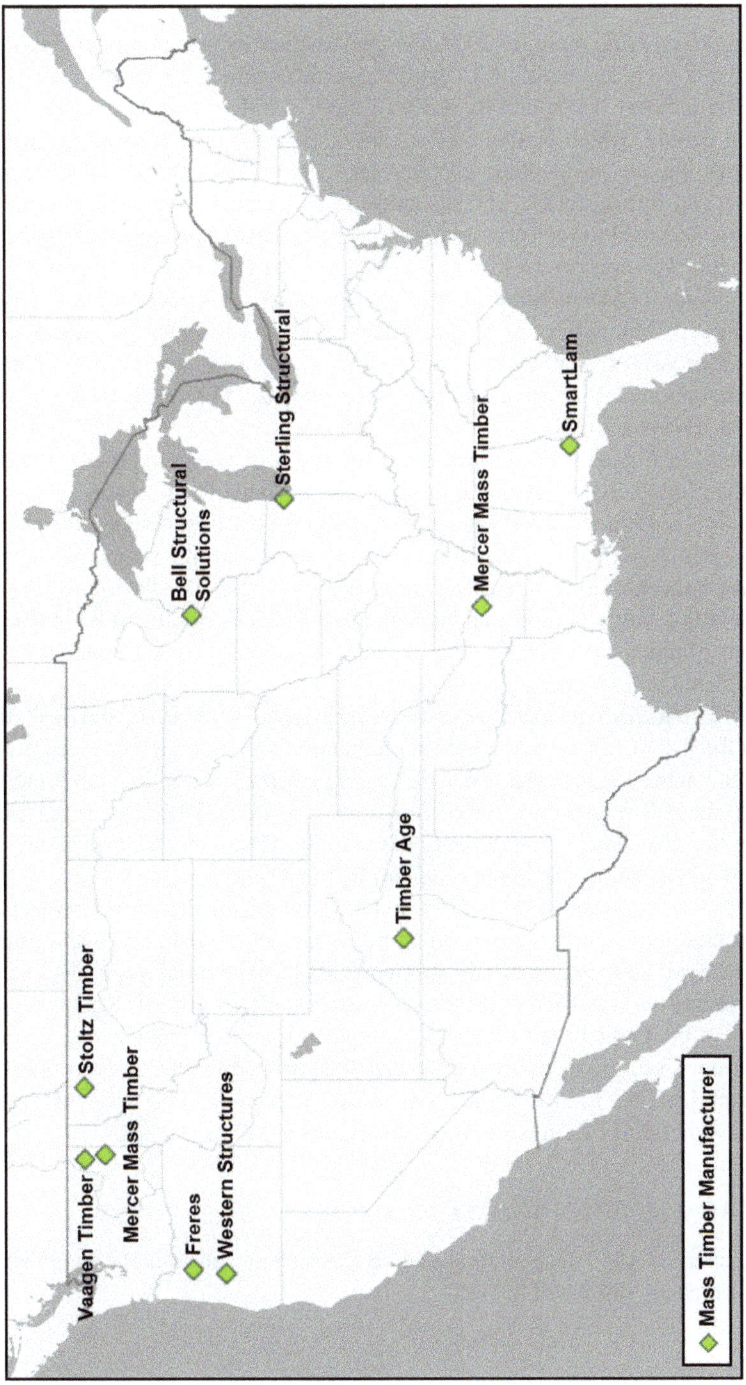

FIGURE 1.16 Current Domestic Mass Timber Manufacturing Locations as of 2024. Courtesy of Amanda Colucci, Ph.D.

1.4 MT IN THE CONSTRUCTION INDUSTRY

According to an AGC report in 2023, the construction industry employs 8 million employees and creates nearly $2.1 trillion worth of structures each year (AGC, 2023). The industry accounted for 4.3% of the U.S. GDP in 2023 and employed 8 million people, which is about 3% of the U.S. workforce. The major sectors in construction are residential, commercial, energy and utilities, institutional, industrial, and infrastructure. MT has emerged as a viable sustainable alternative to concrete and steel for construction of high-performance large and tall buildings. In 2021, the ICC approved seventeen changes to the IBC and International Fire Code, allowing for MT buildings up to eighteen stories. This paved the way for the proliferation of MT construction across the country. As of June 2021, there were 1,169 MT projects (constructed, under construction, or in design) in the United States. That number doubled within three years with 2,253 projects by September 2024 (WoodWorks, 2025).

The map in Figure 1.17 outlines the distribution of commercial MT projects across the United States. Importantly, MT can be built in any geography. The technology is not limited to any type of location as almost all states have completed a MT project. While every state is home to a commercial MT project, it is clear that heavy concentrations of projects are in specific states. The Pacific Northwest has been both a major producer and consumer of MT. The Southeast has begun to emerge as a major contributor to the industry as indicated by the manufacturing and completed project maps.

The 2021 IBC included three new construction types (IV-A, IV-B, and IV-C) that allowed the use of MT or non-combustible materials in buildings up to eighteen, twelve, and nine stories, respectively. The code permitted MT ceilings and integral beams to have 20% exposure for construction type IV-B. This was increased to 100% in the 2024 IBC (a significant change). The IBC with MT provisions has now been adopted either in whole or in part by most states.

Despite the significant spread of MT and the growing number of MT projects, it is important to note that it is a very small percentage of the industry. According to the U.S. Energy Information Administration's 2018 Commercial Buildings Energy Consumption Survey (CBECS), there were an estimated 5.9 million U.S. commercial buildings in the United States in 2018. Annual construction volume statistics tell the same story. For instance, between March 2023 and September 2024, there were 500 MT projects built in the United States. By comparison, an average of 81,900 commercial buildings were built every year between 2000 and 2018.

1.4.1 MT as an Alternative to Concrete and Steel

MT has emerged as a viable alternative to both concrete and steel for a number of reasons which are discussed below.

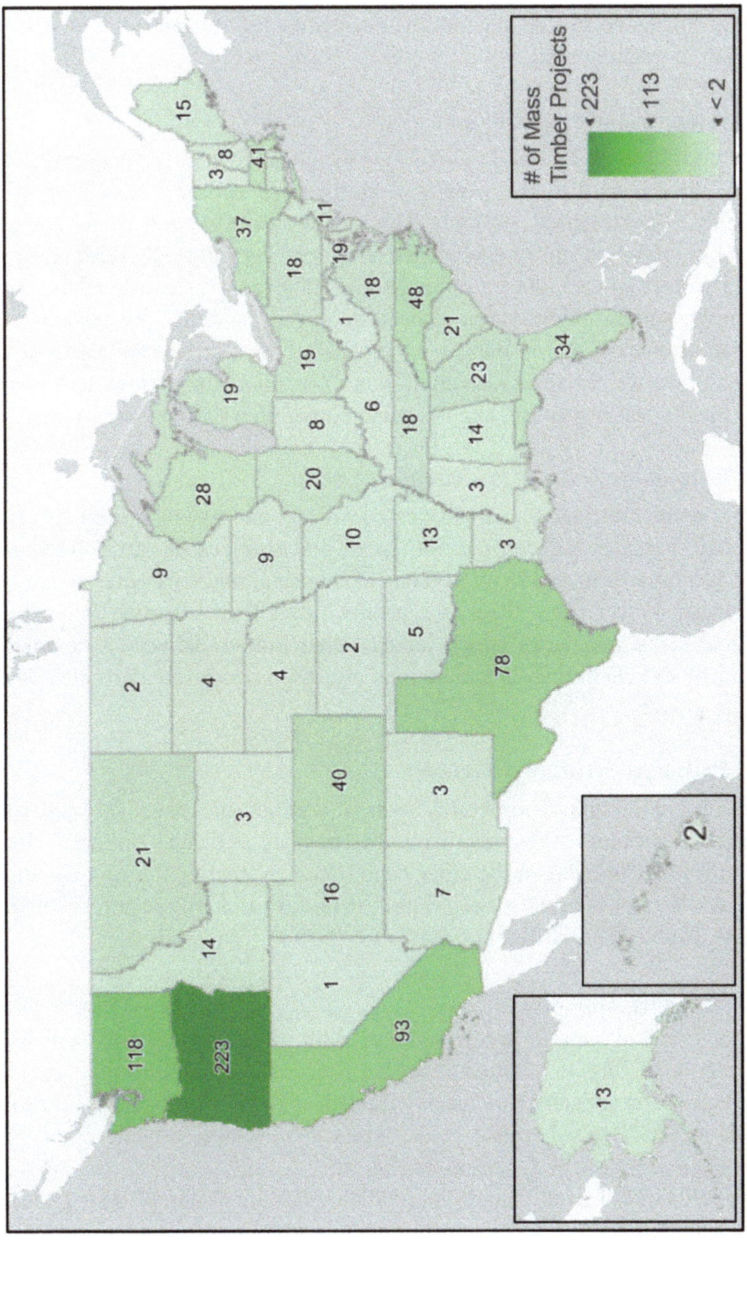

FIGURE 1.17 Mass Timber Projects across the US. (Courtesy of Amanda Colucci, Ph.D).

1.4.1.1 MT is a Renewable Resource

The sustainability argument for MT is overwhelming. As opposed to both concrete and steel, MT is a renewable resource that also acts as a carbon bank and is less carbon intensive to manufacture. The sustainability chapter delves into this comparison in greater detail.

1.4.1.2 Strength-to-Weight Ratio

MT has a high strength-to-weight ratio. They weigh significantly less than concrete and steel (about one-fifth the weight of comparable concrete structure), while maintaining high structural strength. This enables construction to be achieved where weight may be a limiting factor for designers. By using MT, designers have flexibility in erecting structures that could not be achieved with traditional materials. For example, vertical additions have been deployed using MT by installing the lighter frames on existing masonry or concrete structures to increase capacity. The high strength-to-weight ratio also translates to smaller foundations that can be easier to install, less costly and can be delivered within a shorter time frame.

1.4.1.3 Time-to-Delivery (Construction Speed)

MT components are typically prefabricated off-site and assembled on-site. This significantly increases the construction speed and shortens the time-to-delivery. Concrete requires form work, reinforcement, pouring, and cure time which take a significantly longer time. Steel is typically faster than concrete, but it relies on concrete slabs and decks which increase that time-to-delivery. Construction speed and other scheduling considerations are discussed in greater detail in the scheduling chapter.

1.4.1.4 Complete Structural System

MT provides a complete structural system without the need for additional attachments or systems. MT components can be used as beams, columns, floors, walls, and roofs. Concrete on the other hand relies on steel reinforcing for tensile strength, while steel often needs a different material (primarily concrete for slabs or decking. Both require different systems for their walls as well.

1.4.1.5 Aesthetic Appeal

MT offers exposed natural wood surfaces with structural systems which are aesthetically appealing. In addition to aesthetic appeal, studies have suggested that there are considerable health benefits of MT, such as biophilic properties, improved indoor air quality (IAQ) because of the absence of volatile organic compounds (VOC), reduced carbon, and thermal insulation.

The benefits of MT over the conventional methods are discussed throughout the text in greater detail.

1.4.2 Market for MT

It has been observed that MT has significantly grown in the United States. This growth in MT adoption can be attributed to a convergence of factors, including enhanced efficiency, reduced construction costs, aesthetics of exposed wood, the improved occupant health and well-being, the reduced jobsite noise, and the inherent sustainability of MT as a building material. It is, as such, an alternative that should be considered during the project conceptualization.

Current data from the existing MT projects in the United States show that a majority of them are Business/Office buildings at 39%. Multi-family at 22%, Assembly at 15%, and Educational at 15% make up a substantial portion of projects (see Figure 1.18). These are areas projected to see increased use of MT in the future.

1.4.3 Other Applications of MT

Demand for MT products extends beyond building structures. This section highlights several examples of other applications for MT, showcasing the material's adaptability and versatility. The growth in these peripheral uses underscores the material's potential in a variety of industries. Additional breakthrough uses are explored in Section 1.6.2.

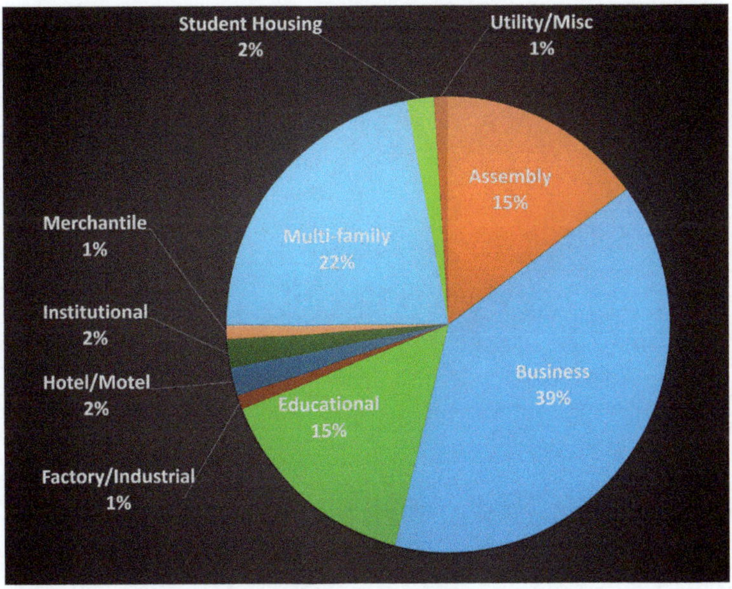

FIGURE 1.18 Mass timber Distribution by Project Type.

1.4.3.1 Industrial Mats

CLT has been used to build temporary access roads as part of site access or ground protection mats as working platforms for industrial applications. While CLT is manufactured using softwoods such as pine, spruce, douglas fir, and larch, traditional industrial mats were made of hardwoods. In 2014, Sterling started the manufacture and use of CLT for this application (see Figure 1.19). Because of its durability and strength, southern yellow pine (SYP) was the species of choice for manufacture of industrial mats.

1.4.3.2 Bridges

MT is gaining traction as an alternative for bridge construction primarily because of its lower environmental impact, strength and durability, aesthetic appeal, high strength-to-weight ratio, fast installation time, and design versatility. For instance, the installation of TCC bridges has increased over recent years. TCC bridges are not a modern application in the United States. During the great depression, steel costs skyrocketed forcing the industry to explore alternative materials for bridge construction. The TCC system emerged as an economical highway bridge solution. Composite slab systems consisting of a NLT slab-type deck with a concrete deck overlay were developed and field tested. In the following years and decades, variations of TCC were explored including the use of glulam beams. This industry sector continues to develop and grow alongside building structure and other uses.

FIGURE 1.19 Industrial Matts (Courtesy of Sterling Solutions LLC).

An example of this application is discussed in *Case Study V: High Line Timber Bridge, New York City, NY* (see Case Study section of this Textbook). The project featured a 260-foot-long warren truss bridge in New York city that is made from sustainably sourced Alaskan yellow cedar glulam beams.

1.4.3.3 Non-Load-Bearing Architectural Applications

MT products are being used for inherent flexibility and adaptability in providing aesthetic applications. Non-load-bearing uses of MT products are often used in conjunction with structural elements. Figure 1.20 demonstrates the design team's choice of using MT for the expansive structural system in conjunction with a non-load-bearing architectural finish application on the interior ceiling. Manufacturers can produce both elements and project teams should consider sourcing interior finish materials from structural suppliers. The project highlight below also uses CLT panels for storefront security barriers that can be raised when the shop is to be opened. Project teams do not typically source architectural finish materials from traditional structural building suppliers, a nuance that should be noted by teams. Sourcing architectural finishes from the structural supplier has multiple benefits, including dollar cost averaging, procurement of structural and finish specified materials, reduced shipping cost, and faster lead times.

FIGURE 1.20 Redmond Senior and Community Center (Courtesy of Freres Engineered Wood).

1.4.3.4 Other Construction Applications

The flexibility of EWPs has driven their adoption in building sectors beyond multi-family, commercial, and higher education facilities. Successful commercial projects are now emerging in warehouses, data centers, distribution centers, and other industrial sectors. Incorporating MT into areas traditionally built with precast concrete allows for faster construction, reduced carbon emissions, and smaller transportation impacts. Industrial center owners have recognized the benefits of using timber products, contributing to the growth of projects in these sectors.

In addition to industrial applications, other sectors are also benefiting from MT structures. Examples include pavilions, amphitheaters, equestrian facilities, art installations, and vertical additions. These projects highlight the versatility and growing demand for MT across various types of developments. Project teams are increasingly seeking ways to incorporate this material due to its aesthetic appeal, environmental benefits, and ease of construction.

1.5 ORGANIZING AND LEADING A MT PROJECT

MT technology is a relatively new way of delivering commercial grade projects, especially in the United States. While popularized in Europe in the 1980s and 1990s, it wasn't until the early 2000s that MT started to be used in the United States on commercial scale projects. Records from WoodWorks indicate that there were 2,253 multi-family, commercial, or institutional MT projects in the design phase, under construction, or already constructed as of September 2024 in the United States. This is a significant increase from the 439 projects in September 2018 and only 26 projects in 2013. While MT is gaining popularity and growing in the United States, it is important for designers and builders to consider best management practices relative to construction of these projects.

To effectively manage MT projects, it is important for one to consider the unique conditions that the technology presents and how the management of those conditions differs from traditional projects. While the core project management principles apply to MT projects, there are distinctive risks and challenges that MT presents, and successful management of these projects must take those considerations into account. Successive chapters in this text will address the distinctive nuances associated with managing MT projects.

The primary objective of project management is to create a sustainable and quality project, on time, within budget, and within the context of a safe work environment. Managing a MT project requires consideration of the unique challenges associated with sustainability of the material, quality of the final product, time (schedule) of delivery, cost (budget) of the project, and safety of the workers and end user. Successive chapters will deal with details of these unique conditions.

1.5.1 Stakeholder Involvement in MT Projects

Because of industry unfamiliarity with the technology, lack of municipal experience, evolving codes and guidelines, and the fact that MT structures heavily rely on pre-manufactured components, there are often multiple stakeholders involved in these projects. Project teams should be aware that an increased number of stakeholders are likely to be involved on MT projects. As the driver of the delivery methods, owners often hire multiple stakeholders to consult and aid in the delivery process, especially when a perceived risk is present. The major stakeholders are developers, designers (multiple structural, architectural, and interior), manufacturers, contractors, and government officials.

As MT projects grow the streamlining of stakeholders will begin, reducing the amount needed. Manufacturers are beginning to offer project management, structural design consultation, and other services that will eliminate the need for multiple parties.

1.5.2 Delivery Methods for MT Projects

Project delivery refers to the process through which a construction project is designed, constructed, and turned over. There are several construction project delivery methods, each with their unique characteristics. Choosing the right project delivery method is key to the success of the project, primarily because the delivery method determines the relationships between the various stakeholders, their roles, how they communicate, and how payments are made. The most common delivery methods in construction are Design-Bid-Build (DBB), also called the traditional delivery method, Design-Build (DB), Construction Manager at Risk (CMAR), Multi-Prime Contracting, and Integrated Project Delivery (IPD).

Unlike the other construction methods (concrete and steel) MT relies heavily on prefabrication of customized components. This necessitates early integration and collaboration with manufacturers, suppliers, designers and constructors, early logistical considerations, and early extensive planning. Because of these needs, MT projects cannot be delivered using the traditional delivery method. Some of the more conventional methods like DB, CMAR, and IPD have been used, albeit with a few modifications.

When selecting a delivery method for any project, the primary consideration is to use a method that facilitates communication, minimizes risk, and enables successful delivery of the project. The following considerations need to be taken for MT projects:

- **Prefabricated Components**: MT elements mostly prefabricated off-site, requiring detailed design with BIM integration to ensure precise manufacturing and smooth assembly on the construction site. These elements call for a collaborative approach to delivery that involves early integration of all project participants.

- **Early Planning and Design Coordination**: Early planning and coordination is vital to ensuring that the right MT components are selected, panel sizes are optimized to minimize waste, and efficient logistical plans are put in place.
- **Early Integration of Manufacturer/Supplier**: Selecting and involving a MT supplier early in the design phase is important to leverage their expertise on material, logistics, and potential cost implications.
- **Logistics and Transportation Considerations**: Working on a supply chain issue early is key to success of a MT project. Considering the means of transportation and transportation routes based on panel size and origin of the materials is crucial. Many projects in the United States have relied on material that is shipped from Europe and Canada. Logistical issues need to be addressed early in the process. These significant lead times also call for early stakeholder integration.

1.6 CURRENT TRENDS AND THE FUTURE OF MT

1.6.1 Current Domestic Industry Trends

Use cases of MT in construction projects are highlighted throughout the text. This section focuses on actual data assimilated relative to international and domestic uses, investments, and other key indicators that showcase the current trends associated with MT.

Firstly, focus will be on completed and/or in progress projects in the United States within the past few years (WoodWorks, 2024):

- **2024**: 2,253 completed multi-family, commercial, or institutional projects
- **2023**: 2,035 completed multi-family, commercial, or institutional projects
- **2022**: 1,300 completed multi-family, commercial, or institutional projects
- **2020**: 784 completed multi-family, commercial, or institutional projects
- **2019**: 600 completed multi-family, commercial, or institutional projects
- **2018**: 439 completed multi-family, commercial, or institutional projects.

Impressively, this represents a 38.6% year over year growth rate in MT project delivery. Sheer project growth should be considered one of the main indicators of strength and future project utilization. Steady growth of projects in the United States is similar to international project explosion. Internationally, MT project numbers are expected to reach over 24,000 by 2034. In terms of dollar value, the MT sales were expected to exceed $1.5 billion (Salhotra, 2023) and upwards of $3 billion by 2030.

In addition to market value and project completion rates, there are several other important indicators to consider relative to the product's trend profile. Increases in federal, state, and organizational funding relative to MT are on the rise. Recent announcements from the federal government include $32 million to fund wood innovations and community wood grants with the specific goal of strengthening and expanding the emerging MT market (Mercer Mass Timber, 2024). Continued investment in the United States Departments of Agriculture's Forest Service Division, through the Wood Innovation Grant in continuing to grow in monies and opportunities for the market segment. Finally, the National Defense Authorization Act (NDAA) passed their fiscal year 2023 budget which included language to establish a pilot program to evaluate effects of MT products in Department of Defense military construction projects. These examples highlight the growing interest and investment in the MT market and should be indicative of the staying power and growth potential of the industry.

Trends in important building attributes are equally important to understand. Several key indicators in the shift of methodologies and building attributes are worth considering. For example, building height is an important building attribute that should be considered in the MT market. Initially, MT projects were constructed in the one to two story range as codes were not conducive, nor were risk appetites aligned with attempting taller MT structures. However, as codes and familiarity with the product has expanded, building heights have increased over time. Prior to 2021 IBC changes, most codes restricted MT building to a maximum heights of five to six stories due to fire safety and code concerns. After the IBC changed classification ratings for timber structures, Type IV-A, IV-B, and IV-C construction types allowed for taller MT buildings, enabling up to 18 stories of construction to take place. Taller MT buildings are on the rise, indicative of the changing nature of code requirements and increased familiarity of the capabilities of the product.

1.6.2 Evolving Uses of MT

Uses of MT in a non-building structure sense are beginning to arise from researchers and construction teams alike. The flexibility of applications of MT is a strength of the product class and this section will highlight some of the more peripheral and experimental uses that are currently being explored. It is difficult to predict all future applications of the product, but interesting work is being done on unique applications that are worthy of being presented here.

1.6.2.1 Blast Shields

In 2017, the USDA Forest Service's Forest Products Laboratory (FPL), WoodWorks, and the Softwood Lumber Board conducted live blasting tests on three, two-story high, CLT structures with promising results. Each of the structures was made with a different grade of CLT and included window and door openings. The results were positive and even exceeded the design expectations, as the structures remained intact and safe to enter, even after the most extreme blast loads (WoodWorks,

2022). CLT's ability to withstand blast loads is due to its multiple laminated layers, which provide residual strength after a blast.

1.6.2.2 Noise Barriers

Most highway corridors adjacent to residential developments require noise barriers to limit sound pollution relative to vehicular traffic. Recent research conducted at Clemson University evaluated the potential for using CLT assemblies for noise barrier applications. This could have an impact on CO_2 emissions generated by eliminating the use of traditional cast concrete panels found in a majority of noise barrier applications (Bothra, 2021). The study determined that CLT panels can be used in lieu of concrete panels: reducing CO_2 emissions and transportation costs, requiring smaller manpower needs, increasing design flexibility, and producing less waste.

1.6.2.3 Wind Turbines

Integrating MT components with other materials can provide opportunities for increased sustainability and efficiency. An example of this includes the integration of manufactured timber into a full-scale wind turbine. Current wind turbines are often scrutinized for high embodied energy costs, difficult recyclability characteristics, and lack of other sustainability considerations. The Swedish company Modvion has constructed a 105-meter-tall tower in the Skara region of Sweden (Aouf, 2024). In this case the company utilized traditional turbine building materials in the rotor blades and generator hub but deployed engineered wood in the tower component. Touting its lightweight to strength ratio, the wood tower enhances the sustainability and cost benefits of existing wind turbine methodologies.

1.6.2.4 Single Family Housing

The United States (among other countries) has been experiencing both a housing availability and affordability crisis for a number of years. MT is being explored as a potential solution to housing issues. Currently, it is tough for single family housing applications to pencil out on the affordability side, but there are massive efforts underway to attack the economic headwinds associated with the product. A company based in Seattle, WA, Green Canopy has published a recent case study on a MT model that touts some serious benefits. A 44% faster construction time relative to traditional methodologies, increased cost control, benefits from indoor manufacturing style construction, up to 6.5 more carbon storage than a stick frame home, and a completion timeline of less than 100 working days.

These examples showcase the flexibility of MT in a broad range of industries and applications. Continued investment and research in engineered timber is bringing exciting results. Project teams are encouraged to think outside of the box during the design phase to look for interesting ways to maximize utilization of the material.

1.7 CHAPTER SUMMARY

The history and application of MT in the built environment is diverse and ever evolving. Since the earliest found examples of wood construction some 476,000 years ago, humans have found new ways to maximize the inherent benefits of wood construction (Spitzer, 2023). The rapidity of utilization over the past ten years has driven new and exciting developments in the timber industry, which was mostly stagnant throughout the last couple of decades. With new technological advances in the building materials industry, construction management teams must be well versed in the pros, cons, and applicability of the material. Clients rely on our expertise in emerging technologies, which this text is poised to provide.

MT construction is unique because it affects every pillar of construction management. This demands extensive effort on behalf of the construction management team to successfully leverage the use of MT. If one pillar of construction management relative to MT is mismanaged, it could mean failure of the project. Such importance is focused on the structure of our buildings as it is critical to project success. Heeding the lessons and resources provided in this text will arm construction teams with the knowledge to manage the complex construction processes associated with MT.

Project management teams must promote flexibility and adaptability as not only are MT technologies deployed, but others. Examples include artificial intelligence, drone technology, autonomous vehicles, and carbon accounting, among others. Construction managers' contractual scope expectations are changing just as rapidly as material and technology advancements. Successful managers will immerse themselves in gaining valuable knowledge in each facet. This chapter provides readers with foundational knowledge in history, development, applications, and market trends associated with MT. Subsequent chapters will provide integral information relative to the pillars of construction management and MT.

1.8 REVIEW QUESTIONS

1. The first patent for CLT applications was developed in which country, and in what year?
    a. USA, 1920
    b. France, 1985
    c. Austria, 1988
    d. Germany, 1971
2. The proliferation of mass timber as a commercial construction material is highlighted by which of the following market indicators?
    a. Increase in manufacturing production
    b. Implementation of mass timber into international code standards
    c. Continued development in research, marketing, and available resources
    d. Investment in new manufacturing facilities
    e. All of the above

3. Considered a major producer of mass timber, this country has not only seen production increases, but is home to over 700 mass timber projects, with 140 currently under construction and/or planned.
   a. Brazil
   b. Argentina
   c. South Africa
   d. **Canada**
   e. Ukraine
4. Current mass timber manufacturing technology has had a positive impact in which of the following key areas (choose multiple)?
   a. **Increased quality control**
   b. **Reduction in needs of manufacturing manpower**
   c. Increased accidents
   d. **Higher levels of flexibility**
   e. Reduction in customization options
5. Future uses of mass timber are broad and exciting. What are some examples of future applications highlighted within the text (choose multiple)?
   a. **Single family tiny homes**
   b. **Wind turbines**
   c. Access mats
   d. Bridges
   e. **Blast shields**
   f. **Noise barriers**
6. The majority of mass timber production currently emanates from what region?
   a. **Central Europe**
   b. South America
   c. North America
   d. S.E. Asia
7. Which class of mass timber products is the most utilized in commercial applications?
   a. DLT
   b. NLT
   c. **CLT**
   d. None of the above
8. Because of the cross-laminations, CLT offers two-way span capabilities and is especially suitable for floors, walls (interior and exterior), elevator shafts, and roofs.
   a. **True**
   b. False
9. What characteristic of mass timber has made it possible to erect structures that could not be achieved with traditional materials such as vertical additions to existing structures which increase capacity?
   a. Renewable aspect of mass timber
   b. **Mass timber's high strength-to-weight ratio**

   c. Mass timber's short time to delivery
   d. Mass timber's aesthetic appeal
10. While mass timber has primarily been used for structural building components, there are alternative applications, such as industrial mats, which are used to build temporary access roads or working platforms for industrial settings. What type of mass timber product is primarily used for these applications?
   a. Glued-laminated timber
   b. Dowel-laminated timber
   c. **Cross-laminated timber**
   d. Nail-laminated timber

REFERENCES

AGC. (2023). Construction Data | Associated General Contractors of America. www.agc.org/learn/construction-data#:~:text=Construction%20is%20a%20major%20contributor,worth%20of%20structures%20each%20year.

AISC. (2025). Steel Design Student Competition | American Institute of Steel Construction. www.aisc.org/education/university-programs/student-design-competition/

Allan, B., & Eaton, D. (2024, June). *The MT roadmap*. The Mass Timber Roadmap. www.fpac.ca/reports/the-mass-timber-roadmap

Aouf, Rima Sabina. (2024, April 5). First full-height timber wind turbine opens in Sweden. Dezeen. www.dezeen.com/2024/01/29/first-full-height-timber-wind-turbine-opens-sweden-design-news/

Bothra, Harsh (2021). Development and promotion of mass timber noise barrier for highways. All Theses. 3681. https://open.clemson.edu/all_theses/3681

Dawson, E., Anderson, R., & Muszynski, L. (2023). *2023 International MT Report*. Forest Business Network.

Kremer, P., Abbasnejad, B., Ahankoob, A., & Wakefield, R. (2024). An Assessment of the Quantifiable Effects for Expressed Timber Biophilic Treatments: A Case for Mass Timber Construction?. *Mass Timber Construction Journal*, 7(1), 31–47. Retrieved from http://journalmtc.com/index.php/mtcj/article/view/40

Mercer Mass Timber. (2024, April 25). Are mass timber buildings the solution to the Affordable Housing Crisis?: Mercer mass timber could mass timber buildings create affordable living spaces? https://mercermasstimber.com/2024/04/25/mass-timber-buildings-affordable-living-spaces/#:~:text=What%20does%20Mass%20Timber%20Bring,a%20Mercer%20Mass%20Timber%20partner

Mirando, A., Onsarigo, L., & Krsteska, I. (2024). Determining the Sustainability of Mass Timber Structures Through Third-Party Sustainability Assessment Systems. *Proceedings of 60th Annual Associated Schools*, 5, 405–413.

Neutral. (2024). A Developer's Guide to Mass Timber | Neutral | Blog. Neutral.us. www.neutral.us/learn/a-developers-guide-to-mass-timber

Roche, D. J. (2023, August 24). Cross-laminated timber (CLT) turns 100. The Architect's Newspaper. www.archpaper.com/2023/08/cross-laminated-timber-clt-turns-100/

Salhotra, P. (2023, October 19). A shift in how we build commercial buildings could be a boon for East Texas. The Texas Tribune. www.texastribune.org/2023/10/19/mass-timber-trend-east-texas-forest/#:~:text=Globally%2C%20the%20number%20of%20mass%20timber%20construction,according%20to%20the%20international%20mass%20timber%20report.

Seagate (2018, November 28). How CLT is manufactured. https://seagatemasstimber.com/how-is-clt-manufactured/

Spitzer, G. (2023, September 22). World's oldest wooden structure defies stone-age stereotypes. NPR. gwww.npr.org/sections/goatsandsoda/2023/09/22/1200975292/worlds-oldest-wooden-structure-defies-stone-age-stereotypes#:~:text=University%20of%20Liverpool-,Archaeologists%20dug%20into%20a%20riverbank%20in%20Zambia%20and%20uncovered%20what,we%20see%20Stone%2DAge%20people.&text=The%20find%20didn't%20look,lying%20crosswise%20over%20another%20lo

Tao, L. (2024, June 3). China unveils groundbreaking mass timber construction in the Landmark Hospital Project. Canada Wood Group. https://canadawood.org/china-unveils-groundbreaking-mass-timber-construction-in-landmark-hospital-project/

WoodWorks. (2022, April 19). Mass timber protective design – woodworks: Wood products council. www.woodworks.org/learn/mass-timber-clt/protective-design/

WoodWorks. (2024, October 3). About WoodWorks. Retrieved from www.woodworks.org/about/

WoodWorks. (2025, March 1). Mass timber projects in design & constructed. Retrieved from www.woodworks.org/resources/mapping-mass-timber/

2 Scheduling Mass Timber Projects

"The construction industry is ripe for disruption. Large projects across asset classes typically take **20 percent longer to finish than scheduled** and are up to 80 percent over budget. Construction productivity has actually declined in some markets since the 1990s; financial returns for contractors are often relatively low—and volatile." (Agarwal et al., 2016)

LEARNING OBJECTIVES

- Understand current scheduling concepts
- Understand critical path considerations on mass timber projects
- Know the differences between scheduling traditional structures versus mass timber
- Understand production rates for mass timber structures.

2.1 OVERVIEW OF CONSTRUCTION SCHEDULING

Accurately forecasting a construction schedule is one of the most challenging tasks for project teams. Scheduling is not just a fundamental aspect of construction management—it is a core competency for project managers. Just as a nurse must be proficient in inserting an IV or a CPA in reading a balance sheet, a project manager must master scheduling. Beyond time management, scheduling impacts quality, sustainability, cost, and safety. Effective scheduling determines the erection sequence, construction methods, and essential resources, including manpower, materials, and equipment.

Schedules are dynamic and fluid, influenced by numerous variables—some within the project team's control, others not. Construction teams must analyze project-specific data to determine the time and resources needed for successful completion. Among the most critical managerial decisions is sequencing construction operations, which directly affects efficiency and overall project

DOI: 10.1201/9781003482543-2

success. Key considerations for schedulers include procurement, manpower availability, subcontractor coordination, and a deep understanding of material properties and processes for critical path items. Addressing these factors allows for the development of accurate and achievable schedules.

Despite advancements in training, technology, and experience, large commercial construction projects take about 20% longer than anticipated. Several factors contribute to this, all of which are explored in this chapter. The primary risks to schedules include global supply chain disruptions, labor shortages, productivity challenges, and inadequate planning or communication. For example, cost-loaded schedule distributions are evolving, requiring significant upfront expenditures to mitigate supply chain risks later in the project. Additionally, the ongoing labor crisis—affecting workforce availability, skill level, and productivity—must be accounted for in scheduling decisions. Project teams that develop expertise in these areas can better manage risks and minimize negative impacts.

One of mass timber's key advantages is its ability to be erected faster and with less manpower compared to traditional structures. However, to fully realize these efficiencies, extensive pre-planning is essential. Project teams must not only develop detailed schedules but also remain adaptable to inevitable delays and changes. While mass timber requires distinct scheduling strategies, like cast-in-place (CIP) concrete and steel structures, it also benefits from many of the same scheduling principles.

In mass timber construction, timing is everything. Success relies on significant preconstruction investment, meticulous coordination, and proactive planning to ensure erection efficiencies are met. This chapter examines essential elements of construction scheduling for mass timber projects, highlighting the unique challenges of procurement, construction, and critical path methodologies specific to mass timber. Since every project is different, scheduling approaches must be tailored accordingly. While focused on mass timber, the scheduling practices outlined in this chapter can benefit all construction projects.

2.1.1 Scheduling Process

Construction scheduling refers to the process of determining the required activities, their respective durations, and sequence of operation with the aim of successfully completing the project. The primary objective of a proper schedule is to outline a plan that, if properly executed, will ensure that a quality project is completed, on time, within budget, while maintaining a safe work environment, and causing the minimum negative impact on the environment. The best schedule is not necessarily the fastest schedule, but the one that meets this primary objective.

Schedules are critical in the construction process because they:

- enable optimization of the project resources (labor, materials, and equipment) maximizing productivity and efficiency
- are required contractually
- increase the likelihood of completing the project on time by proactively identifying and addressing issues that could cause delays

- are a great communication tool that ensures all stakeholders are on the same page. However, one must ensure that the schedule is well prepared and a scheduling technique that all participants can understand is used
- are the key tool used by management to facilitate better coordination between different trades, subcontractors, suppliers, and vendors
- enable management to proactively identify risks, then generate and implement mitigation measures
- allow real-time tracking of the project progress alongside the planned timeline. Deviations from the original plan can be quickly discovered and timely adjustments made
- allow real-time tracking of the project costs alongside the budgeted cost. Cost variances that are generated in a timely fashion and measures taken to ensure that the project is completed within budget
- enable informed decision making by the owner and contractor when changes to the project are needed or desired
- form part of the business record and can aid in preparing claims and proving delay damages.

The scheduling process comprises several steps:

Step 1: Breakdown of the Work

This is typically done through the creation of a work break structure (WBS). A WBS is a visual tool used to progressively break down complex projects into individual components. When applied to scheduling, the WBS further breaks down these components into specific activities (see Figure 2.1).

For projects utilizing mass timber as the primary structural element, it is advisable for teams to allocate this work as a Level 2 work package. This approach facilitates a concentrated allocation of resources and attention to the key structural components of the project.

Step 2: Determining and Assigning Resources and Durations

Once the individual activities have been identified, durations and resources can be assigned. There are several ways that durations are determined.

- For self-performed work, historical production rates can be used to determine the duration. If historical data are unavailable, estimates from nationally published manuals can be used to determine the productivity for specific crews. Input from subcontractors is integral to determining time allocations for specific tasks. Production rate refers to the units of work produced by a unit of equipment or person in a specified unit of time (often 1 hour or 1 day). If the quantity of work to be completed is known (from a quantity takeoff), the overall duration can be determined.

Activity Duration = Quantity/Production Rate

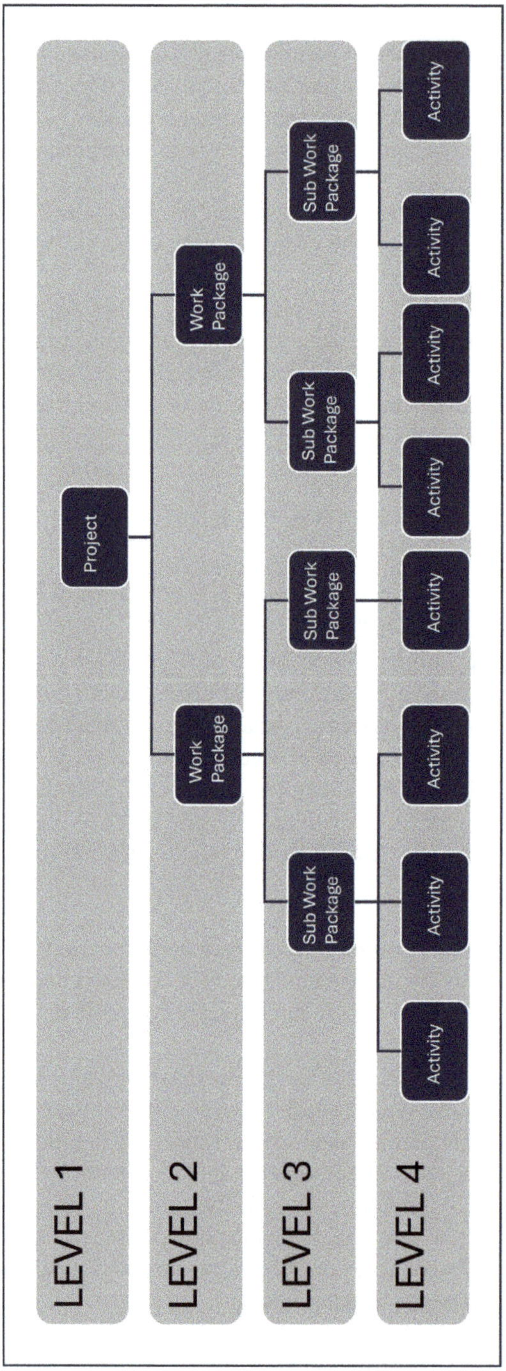

FIGURE 2.1 Work Breakdown Structure.

- As previously mentioned, subcontractors and tradespeople can provide valuable insights into the duration of the tasks they will perform. After pre-qualifying the subcontractor, a request can be made for an estimate of the time required to complete their portion of the work. On mass timber projects, it is crucial to engage in thorough discussions with subcontractors responsible for the structural work package regarding activity durations. These durations are likely to differ from those in traditional construction due to the unique nature of mass timber. Additionally, it is important to assess the subcontractor's comfort level with these changes to ensure smooth project execution.
- In cases where project or activity durations have been predetermined due to specific, unique project conditions, it is essential to allocate resources that are aligned with and capable of meeting those predetermined timelines. Resources include cost, equipment, manpower, and materials. The amount and level of resources included in the schedule will depend on the desired level of information.

Step 3: Determining Relationships and Logical Sequencing

Sequencing and relationships refer to the order in which activities must be executed. Some tasks cannot begin until others are completed, while others can only start after certain tasks have started, and some may occur simultaneously. There are four primary types of relationships between activities (Figure 2.2):

- Finish to Start: This relationship is used when a predecessor activity must be completed before the successor activity can begin.

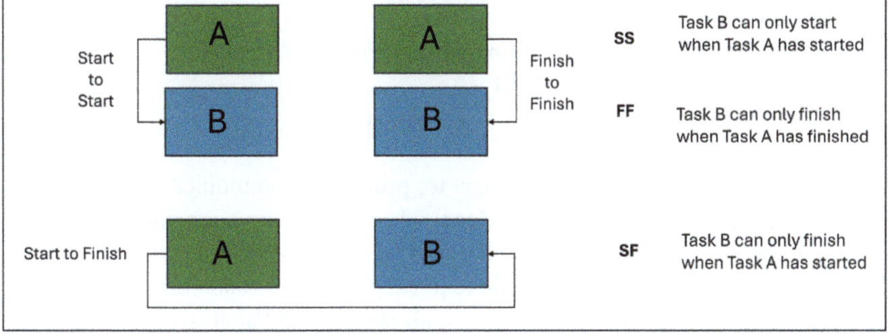

FIGURE 2.2 Activity Relationship Types.

- Start to Start: This relationship applies to activities that can begin simultaneously and run concurrently.
- Finish to Finish: This is typically used when two activities need to finish at the same time, though they may not necessarily start together.
- Start to Finish: This is a rare relationship, but it can be used when an activity must be completed by a specific time after its predecessor has started.

When assigning relationships, the scheduler must carefully consider the construction process, integrating appropriate lags and leads where needed to ensure a logical and practical schedule. Lags represent the time delay between project activities. They are used to account for physical constraints, such as concrete curing time, and to accommodate resource availability, ensuring that tasks are scheduled realistically and efficiently. Leads, on the other hand, refer to the amount of time that can be advanced before an activity begins, allowing for earlier initiation and helping to prevent delays or optimize the overall project schedule. For projects involving mass timber, teams should pay special attention to the relationships of activities immediately following the structural installation sequence. This topic will be explored in greater detail in subsequent sections.

Step 4: Inputting Data to Develop Schedule

This process involves plotting tasks according to the established relationships and sequencing. As mentioned earlier, selecting the appropriate scheduling technique is crucial. Once the data are inputted, the scheduler should carefully review it to ensure accuracy and feasibility. For a more detailed description of the various scheduling techniques, refer to a construction scheduling textbook, as only an overview of these methods is provided in the following section.

2.1.2 Scheduling Techniques

There are several techniques used to communicate a project's schedule including, but not limited to:

- gantt/bar chart,
- critical path method (CPM),
- line of balance/linear scheduling method (LSM),
- last planner system (LPS), and
- program evaluation and review technique (PERT).

Because a schedule is a tool that we use to, primarily, communicate the details of what is being done, when it is being done, where it is being done, and by whom it is being done, management must use a technique that meets that need. For instance, CPM is a powerful tool that is flexible (can adapt to simple and complex projects), can help determine the effects of changes, allows different analysis including cost-benefit analysis, and easily identifies critical activities for management to focus on. On the other hand, one needs training on CPM to be an effective user. While it may be the right tool when communicating with a construction team that has been trained on CPM, it may not be the right tool for an untrained eye, especially inexperienced owners or other partners that don't have training on CPM.

The nature of the project is another key factor to consider when selecting a scheduling tool to use. For instance, line of balance/LSM is best used on projects of a repetitive nature such as multi-family housing, high rise office or apartment buildings, pipelines, and highways. Other projects that are not repetitive in nature like schools and hospitals may not be a good fit for LSM.

It is not unusual to use multiple scheduling techniques on a single project. Modern scheduling softwares have the capability of generating multiple and different types of schedules that are customizable for the respective audience. It is recommended that the project team determine the scheduling technique(s) that work best for the project and project participants.

2.2 SCHEDULING MASS TIMBER PROJECTS

2.2.1 Scheduling the Development of a Mass Timber Project

As outlined in the previous chapter, the development of a mass timber project differs from that of traditional construction methods, requiring earlier and more integrated collaboration among stakeholders. Notably, the construction team must be involved early in the design phase, ensuring a clear understanding of timelines throughout the development process. Several critical phases must be carefully considered by the construction, development, and design teams before construction activities begin. Like all other projects, but more critically in mass timber projects, early development stages present the greatest opportunity for construction managers to positively influence the project schedule and cost as shown in Figure 2.3.

Seagate Mass Timber Inc., one of the largest mass timber constructors, provides an outline of several key stages that teams should consider when developing a project of this type:

Phase 01: Selecting the right Property and Architect

The property characteristics including location, shape, orientation, and type are the primary factors to consider when contemplating utilizing mass timber. The architect and manufacturer will work with the team to maximize site utilization and meet the demands of the property.

Phase 02: Engage a Structural Engineer

Procuring a structural engineer early in the design phase is crucial, especially compared to traditional structures. This early engagement should generate value by setting a budget, schedule and procurement plan, and align team leadership to help through the design and manufacturing stages.

Phase 03: Permitting

Since mass timber structures often undergo greater scrutiny from building departments, the permitting process should begin earlier than with

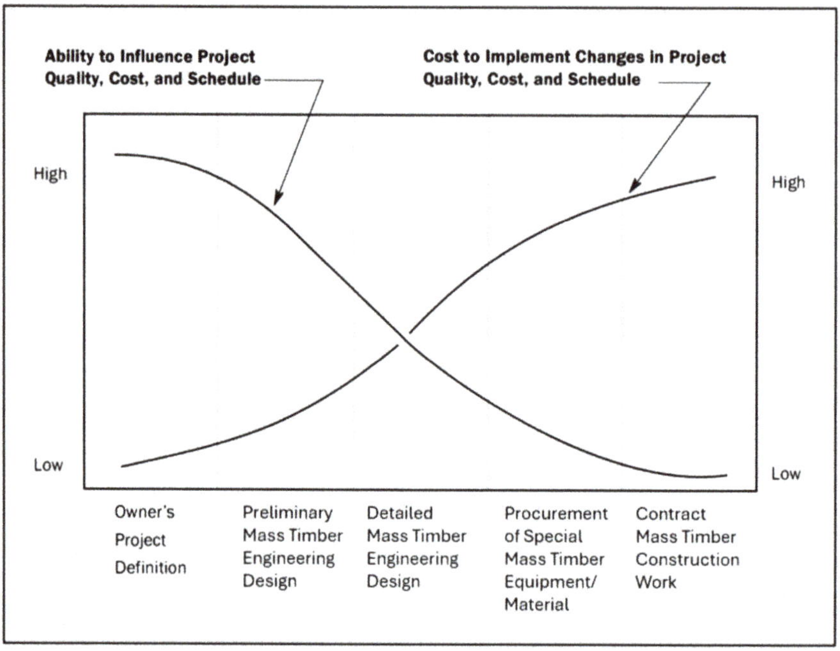

FIGURE 2.3 Ability to Influence Project and Corresponding Cost Implications.

traditional structures. The structural contractor, manufacturer, and architect play key roles in securing timely approvals from the local municipality. This stage is critical to keeping the project on schedule and meeting projected timelines.

Phase 04: Manufacturing

Once approvals have been granted from the engineer and local municipality, the manufacturing process can begin. Project teams must actively communicate and monitor the manufacturing process to ensure that the specific sequence of material delivery and schedule is met. During this phase, pricing, dimensions, and material options are assessed and selected to align with the project's requirements.

Phase 05: Installation

In this phase, the time and effort invested in the previous stages culminate in execution. When significant attention has been given to design accuracy, installation can be highly efficient, particularly since all elements are prefabricated. However, if design changes are necessary, they must be addressed during the manufacturing stage, as the impact of design changes in the construction phase is often more significant on mass timber structures than in traditional construction. The benefits of thorough pre-planning should be fully realized during the erection phase.

The stages outlined above set the foundation for the following sections that discuss in-depth construction scheduling considerations. Since pre-construction investment is critical for constructors to consider, an overall phasing plan was presented.

2.2.2 Mass Timber Lead Times

In construction, lead time refers to the period between initiating procurement (such as issuing a purchase order) and the actual delivery of materials, equipment, or components to the job site. Lead times can include administrative processes such as executing purchase orders, making down payments, obtaining submittal approvals, and securing permits. Several factors influence lead times, including product availability, supply chain efficiency, project design complexity, and the supplier's location relative to the construction site. Additional variables—such as market conditions, labor strikes, holidays, weather, and administrative approvals—can further impact delivery schedules. Construction teams must carefully monitor and account for these timelines when developing and managing project schedules. Accurately quantifying lead times for mass timber components is challenging due to various factors, including global supply chains, design specifications, and manufacturing capacities. However, this section provides a range of expected lead times based on project examples, offering valuable insights into mass timber procurement, key factors affecting lead times, and strategies to mitigate delays.

For mass timber projects, discussing lead times with the selected manufacturer is essential, as production and delivery timelines can vary significantly. If materials are sourced from overseas, extended transportation timelines should be anticipated. For example, the large-scale Ascent project required six months for mass timber prefabrication and delivery due to international sourcing (see Case Study II: Ascent). The complexities of the pre-construction process during this period highlight the importance of early collaboration to prevent schedule delays.

Beyond mass timber components, teams must also consider the procurement and production of fasteners, connectors, and other essential materials needed for structural assembly (Beck, 2023). These typically include steel connection members, plywood splines, and glulam framework. Given the high precision of manufactured timber components, associated hardware must be engineered to exact tolerances to facilitate seamless assembly and efficient erection.

2.2.3 Sequence of Construction Operations on Mass Timber Projects

On mass timber projects, schedule efficiencies can be significant when compared to traditional methodologies and must be integrated into the scheduling process. Failure to incorporate these efficiencies may hinder subcontractors, suppliers, and other stakeholders from executing their tasks efficiently. Compared to traditional concrete and steel construction, mass timber structures offer an increased erection speed of approximately 15–30% (Corgan, 2023). These efficiencies must be accounted for across all construction phases. For example, Figure 2.4 illustrates

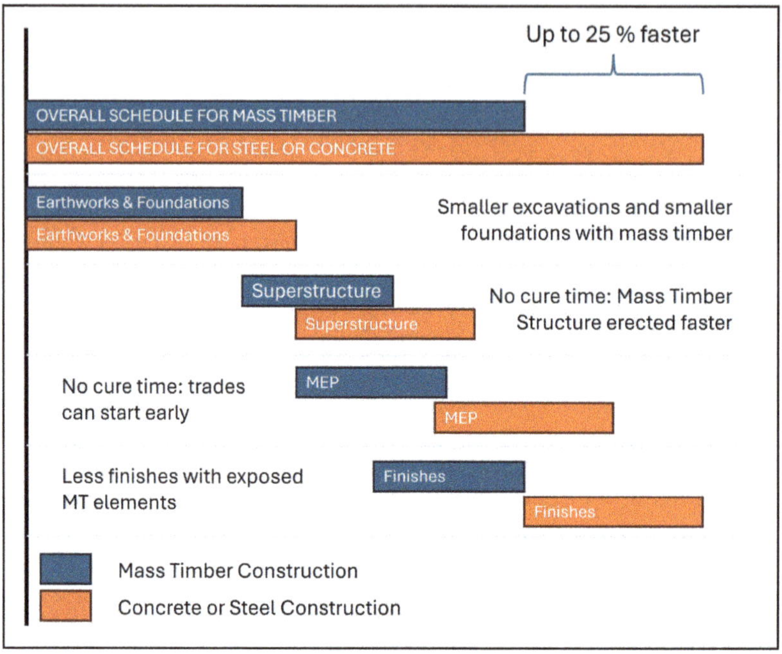

FIGURE 2.4 MT Schedule Comparison with Concrete and Steel.

necessary schedule adjustments, highlighting a predicted 25% reduction in overall construction time. This requires critical line items to start earlier, impacting successor activities that must be realigned to reflect the accelerated erection timeline.

The prioritization of procurement line items is also crucial. Mass timber projects demand extensive pre-planning, particularly regarding the procurement of prefabricated materials. A phased scheduling approach should be adopted, breaking the project into manageable stages: foundational work, structural installation, building enclosure, and finishing. Additionally, continuous evaluation of the critical path is essential to maintain progress and adjust timelines as needed.

Sequence optimization ensures that subcontractors are efficiently realigned throughout the project. As illustrated in Figure 2.5, mass timber construction can have a significant impact on subcontractor sequencing. Trades such as framing, mechanical, electrical, plumbing, and fire protection (MEPFP) can begin installation immediately after erection, unlike CIP projects that require curing and shoring removal before subsequent work can proceed.

Weather conditions must also be considered when scheduling mass timber projects. Being hygroscopic, mass timber absorbs moisture, which can cause swelling, cracking, warping, and discoloration. To help mitigate these risks, schedulers should plan for delivery and installation during more environmentally

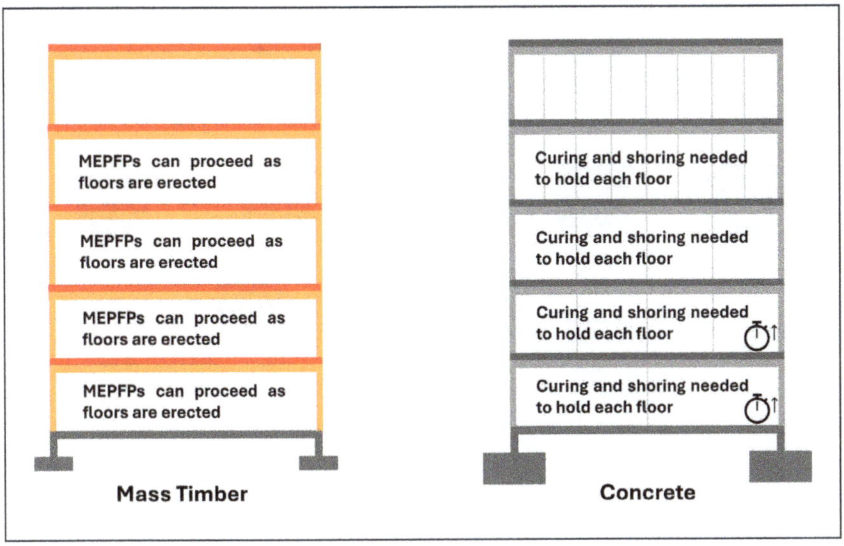

FIGURE 2.5 Scheduling Trades on Mass Timber Projects versus Concrete.

conducive periods (such as minimal rainfall/snowfall) and aim to enclose the structure as quickly as possible. Suggested protective measures, such as temporary weather protection and waterproof membranes, are detailed in the water management plan section within the quality chapter.

2.2.4 Critical Path on Mass Timber Projects

Foundational scheduling knowledge is essential for understanding and leading a mass timber project. This section defines one of the most critical methodologies of construction scheduling: the CPM. Used by project teams and taught in nearly every higher education construction management program, CPM helps manage resources effectively, improve planning, avoid bottlenecks, and prioritize tasks. The development of the CPM is often credited to two major projects in the mid-1950s: the Polaris missile project for the U.S. Navy and a joint effort between DuPont and Remington Rand. Since its inception, CPM has revolutionized project management, particularly in the construction industry, by significantly enhancing efficiency. By providing a systematic approach to scheduling and identifying critical tasks, CPM allows teams to optimize timelines, allocate resources more effectively, and minimize delays, ultimately improving overall project delivery and performance.

The critical path represents the longest sequence of dependent activities required to complete a project on schedule (Figure 2.6). Any delay in a critical path activity extends the project's completion date. Simply put, a critical path activity has no float. If its duration increases by one day, the project completion

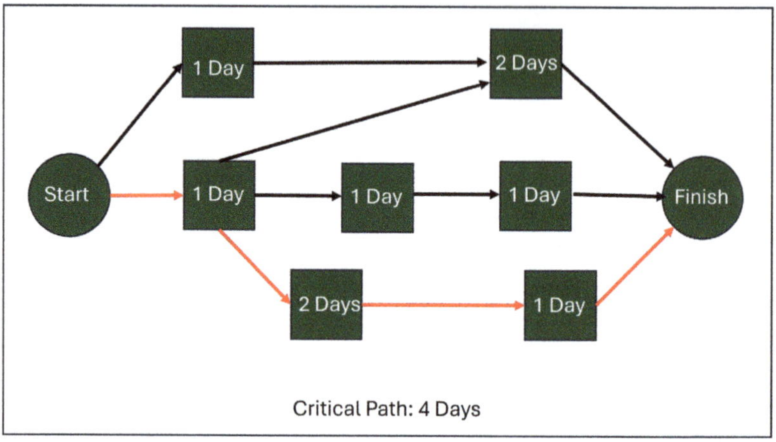

FIGURE 2.6 Critical Path.

date extends by one day. Conversely, reducing its duration advances the project's completion date. Project teams are contractually bound to end dates, and exceeding these deadlines can lead to severe penalties, depending on the contract type. This makes critical path activities exceptionally valuable in construction scheduling. On most commercial construction projects, key activities typically fall on the critical path, including site layout, excavation and foundations, structural framing, MEPFP, roofing, finishes, and final inspections. Mass timber projects follow the same pattern; however, given that their structural components are prefabricated for faster installation, scheduling must carefully account for prefabrication timelines, logistical networks, and jobsite sequencing to optimize productivity.

The critical path varies based on project type, location, material selection, and other site-specific factors, necessitating a thorough analysis using the six steps of the CPM. The six key steps in the CPM are:

1. **Identifying Construction Tasks**—Listing all essential activities required to complete the project.
2. **Determining Task Dependencies**—Establishing predecessor and successor relationships to define the sequence of work.
3. **Creating a Visual Diagram**—Developing a network diagram or flowchart to map out task relationships.
4. **Estimating Activity Durations**—Assigning realistic timeframes to each task based on historical data, expert judgment, or productivity rates.
5. **Calculating the Critical Path**—Identifying the longest sequence of dependent tasks, which determines the shortest possible project duration.
6. **Revising and Optimizing the Schedule**—Continuously updating the plan to accommodate changes, optimize efficiency, and mitigate delays.

This structured approach is particularly valuable for large-scale commercial construction projects, where numerous interdependent tasks require precise coordination. By identifying the critical path, project managers can focus on key activities that directly impact the overall timeline, ensuring efficient resource allocation and proactive risk management.

Notably, project teams with experience in managing a project's critical path are more valuable than those with experience solely in a specific building type. For instance, a project manager who has recently completed a post-tensioned concrete parking garage would be well-suited for another project with a post-tensioned concrete structure, even if the building type differs, such as a higher education facility. Aligning expertise with the primary critical path components is essential for project success.

The significance of structural elements in commercial construction timelines cannot be overstated. Structural frameworks exert a substantial influence on project schedules and must be managed effectively. Given that mass timber technology directly pertains to structural framing, understanding its critical path implications is crucial. Project teams must develop expertise in evaluating and managing the critical path in mass timber projects to ensure successful scheduling and project completion.

2.3 DRIVERS OF EFFICIENCY

2.3.1 Manpower and Mass Timber Structural Erection Efficiency

Access to skilled construction trades is becoming increasingly challenging and competitive, with labor shortages particularly acute in highly specialized and unionized sectors. Trades such as steel erection, masonry, electrical work, and surveying are experiencing a rising demand coupled with a dwindling supply of new entrants. This growing gap makes the recruitment and retention of skilled labor a pressing challenge for the industry, necessitating proactive workforce development strategies and investment in training programs.

Mega projects have further strained municipal labor pools, exacerbating shortages amid a decade-long construction boom. For example, large-scale developments like the Shell Cracker Plant in Pittsburgh, PA, and the Intel Corporation Chip Factory in Columbus, OH, have drawn skilled trades from surrounding municipalities, states, and even countries by offering premium wages. While beneficial for those projects, this migration has intensified labor shortages elsewhere, forcing project teams to proactively consider workforce availability when scheduling construction projects. To mitigate these challenges, project teams must require subcontractors to commit to manpower-loaded schedules within contracts and ensure that realistic timelines are established in response to labor market fluctuations. Transparent communication regarding workforce availability is paramount to project success.

Mass timber structures introduce additional considerations beyond traditional structural systems that impact labor requirements, workflows, and project durations.

Due to the prefabricated nature of mass timber components, significantly less labor is required for installation compared to conventional materials. Mass timber crews typically consist of fewer than half the workers needed for a concrete crew, and the superstructure can be erected up to 25% faster than a comparable concrete structure. Steel structures often require heavily reinforced concrete slabs, necessitating coordination between multiple materials and installation teams, which can impact efficiency. Similarly, concrete structures require formwork and steel reinforcement before pouring, adding additional steps that slow down the construction process. In contrast, mass timber offers a fully prefabricated structural system, manufactured off-site for streamlined on-site assembly. Numerous case studies have demonstrated that mass timber construction achieves higher productivity rates than traditional concrete and steel systems while requiring significantly smaller crew sizes.

A peer-reviewed study (Mirando and Onsarigo, 2022) found stark labor differentials between mass timber and traditional CIP concrete (Table 2.1). When developing manpower-loaded schedules for mass timber projects, these labor efficiencies must be factored in to optimize productivity and resource planning. Leveraging such data-driven insights ensures a more accurate assessment of labor needs, reducing the risk of delays and workforce shortages.

This study highlighted significant labor efficiencies when comparing a large-scale commercial mass timber structure to a comparable CIP concrete building. The mass timber structure required approximately seven skilled carpenters per day, whereas the CIP concrete structure demanded an estimated 18 workers, including concrete laborers, finishers, foremen, pump operators, and carpenters. Not only did the mass timber project require fewer workers, but it also benefited from lower labor costs due to a reduced reliance on specialized trades. Remarkably, despite the smaller crew size, the mass timber team installed 2,322.5 square feet of structure per day, surpassing the 1,824.9 square feet per day achieved with the CIP concrete team.

TABLE 2.1
Mass timber v concrete production rates

	Mass timber	Concrete
Total zone floor area (square feet)	76,644	76,644
Floor levels	6	6
Square feet per level	12,774	12,774
Total working days	33	42
Total man-days	218	756
Total man-hours	**1,744**	**6,048**
Manpower average	**6.6**	**18.0**
Square feet per day	2,322.5	1,824.9
Square feet per man-day	351.6	101.4

While the data presented here were collected after the mass timber team had overcome the initial learning curve, it is important to recognize that productivity on mass timber projects is often impacted during the early phases. Inexperienced installation teams typically face supply chain and installation challenges, which can hinder efficiency. This is a common hurdle when working with a new material and highlights the critical importance of supply chain management, thorough pre-planning, and workforce training to optimize productivity and ensure smooth project execution.

Another compelling example is the San Jacinto College project, completed by Tellepsen builders, Tellepsen compares past steel projects and showcases that they were able to erect a 122,000 square foot structure in a similar time frame to a steel structure, with only a six-man crew. This results in reduced costs, increased safety, and more room for other trades to complete their work. Their experience reinforced the unique advantages of mass timber construction, particularly in speed, safety, and labor efficiency.

Beyond these highlighted cases, there are other project examples that further illustrate the efficiency gains achieved through mass timber construction.

The rapid erection of structural components in mass timber buildings not only shortens overall project timelines but also alters traditional construction sequencing. As a result, subcontractors' schedules must be adjusted accordingly. Some mass timber projects have stalled post-erection because construction teams underestimated the speed of assembly, leaving subcontractors unprepared to begin their work. To prevent delays, schedulers must account for mass timber's accelerated installation and ensure MEPFP and framing contractors are ready to proceed immediately after the structure is complete.

2.3.2 Ease of Fastener Installation

Fasteners play a crucial role in construction projects, as they are essential for securing various components to the structural system. These components often include fire suppression systems, HVAC ductwork, electrical cables and conduits, water supply pipes, and wastewater systems. The installation of these systems is critical to maintaining and enhancing project schedules, as they are typically key elements on the critical path. Their installation begins promptly after the structure has been erected, making the reliability and efficiency of fastening methods vital to overall project success. Proper selection and use of fasteners help ensure structural integrity, streamline installation, and prevent delays.

Material type is one of the factors that has a major impact on fastener installation efficiency. Fastener installation efficiency in mass timber is significantly higher than installation in steel and concrete. Fastener installation in mass timber is two to three times faster than installation in steel. Heavy-duty fasteners can be installed over ten times faster in mass timber than in concrete. For the light-duty applications, the fasteners can be installed over five times faster in mass timber than in reinforced concrete. This disparity can be explained by the fact that fastener installation in concrete requires a dual-motion process: pre-drilling first and then installing the fastener. By contrast, installation in mass timber utilizes a single-motion process. In the mass timber application, the installer only needs one tool (typically an impact driver). But the installer in concrete will typically require two tools: a rotary hammer drill, that can weigh up to 15 pounds, for pre-drilling and an impact driver. While the efficiency on mass timber projects is clear, there are ergonomic implications related to mass timber.

A recent study quantified and projected the impact of increased efficiency on a large-scale mass timber project. In their comparison, Onsarigo and Mirando (2023) identified time savings of approximately 600 man-hours across MEPFP trades on a nine-story, 512,000-square-foot mixed-use commercial development. Project teams are encouraged to leverage these data when coordinating with subcontractors to ensure that manpower reductions and cost savings are accurately reflected in project schedules. Effective communication of these efficiency gains can lead to more precise planning, optimized labor allocation, and improved overall project outcomes.

2.3.3 Smaller Concrete Foundations

One often overlooked advantage of mass timber structures, particularly for those new to the field, is how their lightweight nature reduces the need for extensive concrete foundations. The lighter mass timber superstructure directly translates into smaller, less labor-intensive foundations, requiring less excavation, reinforcement steel, and concrete. While concrete may still be necessary for certain structural elements, such as shear walls, elevator shafts, and retaining walls, the overall demand is significantly lower. This reduction in concrete usage not only lowers costs but also minimizes construction traffic and shortens the critical path for foundation work. To maximize these benefits, it is essential to reflect the reduced foundation requirements in project estimates, adjusting both the schedule and cubic yardage calculations to account for the efficiencies gained.

2.3.4 Subcontractor Alignment

Unlike steel and concrete structures, which require extensive detailing, formwork, and a completed and cured concrete slab for a safe working surface, mass timber allows work to commence immediately after mass timber panels are installed. This advantage enables teams to begin overhead MEPFP rough-ins and framing much earlier, accelerating overall project timelines.

Subcontractors working on mass timber structures benefit from open floor plans with limited obstructions, allowing for greater efficiency. The speed of mass timber erection directly impacts follow-up trades. If sequencing is not carefully managed, MEPFP trades may struggle to keep pace, leading to bottlenecks that negate mass timber's efficiency. Mass timber significantly alters traditional sequencing. Unlike CIP concrete or steel, mass timber eliminates long waiting periods, allowing MEPFP trades to begin work almost immediately. For instance, CIP concrete follows a seven-day cycle—reinforcement, forming, shoring, pouring, and stripping—all delaying MEPFP access. Additionally, shoring remains in place for at least another week, preventing successive trades from accessing lower floors for over two weeks after pouring. This is not the case with mass timber as these subsequent trades can proceed with work soon after the structural components have been erected.

MEPFP trades also gain advantages from pre-drilled penetrations, easy-access fastener installation, and precise system alignment. Data show clear time savings not only in structural installation but also in MEPFP integration. Several factors contribute to these efficiencies:

- **Pre-Drilled MEPFP Penetrations:** Mass timber panels for walls, roofs and floors, and mass timber beams arrive with pre-drilled penetrations, reducing layout and drilling time. This efficiency driver is further expounded in the next section.
- **Access for MEPFP Systems Installation:** Early and easy access with minimal obstructions (like shores and reshores) allow subcontractors to install prefabricated components more efficiently.
- **Simplified Modifications:** Mass timber allows for easier and safer routing changes compared to concrete or steel. Drilling through mass timber elements to accommodate rerouting of MEPFP systems is significantly more straightforward, reducing labor time and complexity. This flexibility enhances adaptability on-site, making modifications more efficient and cost-effective.
- **Readily Available Anchoring Points:** Unlike steel and concrete, which require pre-drilling and complex tie-off points, solid wood provides a more adaptable and efficient medium for securing systems.

It is important to note that these efficiencies also scale with project size, requiring teams to carefully schedule subsequent trades, especially on larger and more complex projects.

Fully capitalizing on the benefits of mass timber construction requires meticulous pre-planning and close coordination with trade partners. In some projects, mass timber structures were erected before MEPFP trades were ready, resulting in costly delays. To avoid such setbacks, rapid structural assembly must be strategically aligned with material deliveries and labor availability to maintain a seamless construction timeline. The San Jacinto College project highlighted the critical role of pre-coordination meetings in ensuring subcontractors are prepared for the accelerated pace of mass timber construction. Proper planning, early communication, and proactive scheduling are essential to maximizing efficiency and preventing workflow disruptions.

2.3.5 Pre-Drilling of MEPFP Penetrations

As discussed in the previous section on subcontractor alignment, a key advantage of mass timber is its ability to be engineered and prefabricated off-site with precut MEPFP penetrations (see Figure 2.7). This eliminates the need for extensive on-site drilling, streamlining installation, and improving overall construction efficiency. Manufacturers can precisely lay out and pre-drill openings in both mass timber panels and engineered beams. Although this approach requires upfront coordination and design time, it significantly benefits trades by eliminating the need for on-site cutting, reducing labor, and minimizing potential errors. Furthermore, if adjustments are necessary, modifying penetrations in mass timber is far easier and more efficient than in concrete or steel, which can be time-consuming and costly.

Mass timber readily accommodates both vertical and horizontal MEPFP openings, with larger penetrations for ducts and shafts engineered to maintain structural integrity. Smaller openings for plumbing and electrical conduits typically have minimal impact on the overall structure. However, considerations for fire resistance and acoustical performance must be addressed. Effective collaboration between design and engineering teams is crucial to ensuring accuracy, compliance with building codes, and optimal performance.

FIGURE 2.7 MEPFP Penetrations.

2.3.5.1 Structural Considerations for Penetrations

As of 2025, there are no standardized guidelines in the United States specifying the required level of reinforcement for penetrations in mass timber panels. However, engineering teams can perform calculations to ensure that PSF load standards are met. Structural modeling is essential for conducting these analyses. A common approach, known as "strip analysis," uses hand calculations and assumes the panels adjacent to the openings behave similarly to beams. While this method provides a general guideline, it is not as precise as more advanced techniques. A more detailed analysis utilizes a finite element model (FEM) to assess the force distribution around openings in panels. Since CLT is orthotropic, meaning it has different structural properties in different directions, structural analysis tools face challenges. However, using an orthotropic thick-plate analysis provides a more accurate engineering model to determine the strength properties of the panels (WoodWorks, 2022).

2.3.5.2 Early Coordination for MEPF Installations

For successful integration of MEPFP systems into Mass Timber, it's essential that MEPFP contractors provide building information modeling (BIM) input early in the design stages. This allows manufacturers to account for elevations, locations, and pipe sizing requirements. Similar to panel penetrations, horizontal MEPFP runs that intersect with beams can also be accommodated, as shown Figure 2.8.

FIGURE 2.8 CLT Ceiling Panels and Beams with Pre-Drilled Penetrations for MEPFP Routing.

These penetrations require more detailed calculations due to the structural nature of beams. However, when properly designed, they save space, ensure accurate layouts, and simplify installation for trades, ultimately enhancing overall project efficiency.

2.3.6 Just-in-Time (JIT) Delivery

JIT delivery is a supply chain strategy focused on delivering products exactly when they are needed for installation. This approach minimizes waste and inventory costs, ensuring materials are available just when required. In Mass Timber projects, implementing JIT requires careful communication and precise delivery planning from the manufacturer regarding component shipments. JIT is particularly beneficial for project teams aiming to reduce weather exposure, increase erection efficiency, and eliminate on-site waste.

To achieve successful JIT delivery, project teams must invest in thorough pre-planning, particularly when managing Mass Timber logistics, deliveries, and installation sequences. Mass Timber components are typically delivered in containers/flatbeds, with beams, columns, and CLT panels clearly tagged for specific installation locations. However, project teams can't always guarantee shipment order due to supply chain disruptions such as customs delays, mislabeling, or labor strikes.

To mitigate these risks, teams should consider utilizing a nearby off-site storage facility. While this adds some construction costs, it improves delivery management by allowing teams to "shake out" shipments and align them with the actual needs of the jobsite. This strategy ensures materials arrive in the optimal sequence for installation, minimizing disruptions. Careful scheduling and clear communication about delivery timing and content are vital to achieving JIT delivery and adhering to lean construction principles.

Often, custom steel connections need to be pre-installed on beams and columns before installation, depending on the design. Installing these connections off-site streamlines the field installation process. Investing in off-site storage and delivery management ultimately yields scheduling benefits, allowing for more efficient assembly of mass timber components. This investment can offer a return on investment (ROI), providing greater flexibility during construction and contributing to overall project efficiency.

2.4 ANTICIPATED DEVELOPMENTS IMPACTING MASS TIMBER PROJECT SCHEDULING

Due to the relative infancy of mass timber, it is essential to anticipate evolving factors as this sector of the construction industry matures. This section outlines key developments expected to significantly influence the scheduling of mass timber projects. While these changes have yet to fully materialize, their impact on the industry is anticipated to become increasingly evident in the near future.

Maturation of the Supply Chain: As demand for mass timber grows, manufacturing capabilities continue to expand. The increasing number of mills and manufacturers, coupled with advancements in production efficiency, is leading to faster milling times, enhanced mass timber capabilities, and broader applications. These evolving factors must be carefully integrated into project schedules to ensure accuracy and efficiency.

Increased Builder Expertise: With more construction teams gaining hands-on experience with mass timber, project timelines are expected to become more efficient. As learning curves diminish, builders will streamline installation processes, reduce delays, and improve overall productivity.

Growth of Prefabrication: The industry is increasingly favoring off-site fabrication due to its consistent and controlled conditions. The integration of mechanical, electrical, and plumbing (MEP) contractors into mass timber prefabrication will have a profound impact on project timelines. This shift necessitates adjustments in scheduling to accommodate new coordination requirements and prefabrication lead times.

Advancements in Engineering and Design Efficiency: As architects and engineers gain more experience with mass timber, the design and approval processes are becoming more streamlined. In the early days of mass timber construction, project teams often navigated a complex web of stakeholders. However, as best practices become more standardized, response times for requests for information (RFIs) are improving, pre-construction workflows are becoming more efficient, and project teams are gaining greater flexibility during construction.

Advanced Technologies: Technology is transforming every aspect of the construction industry, with innovations like artificial intelligence (AI) increasingly being integrated into manufacturing, design, and construction processes. As these advancements continue to evolve, their impact on project scheduling and overall delivery timelines is expected to grow significantly.

By recognizing and adapting to these industry shifts, project teams can proactively optimize their schedules, reducing inefficiencies and maximizing the benefits of mass timber construction.

2.5 CHAPTER SUMMARY

Time is money, particularly in commercial development projects where every day of delay can have significant financial consequences. In many cases, project contracts explicitly outline liquidated damages, attaching a direct dollar amount to each day of delay. Mass timber is frequently promoted as a solution to meet the demands of aggressive construction schedules. However, many projects fail to realize its potential due to insufficient pre-planning and a lack of understanding of the unique nuances involved in working with mass timber. By leveraging the insights provided in this text, teams can maximize the efficiencies that mass timber offers.

In summary, when mass timber projects are properly planned and executed, they not only result in faster construction timelines but also require fewer workers compared to traditional structures. This is due not only to the material's inherent properties but also to the construction methods that are tailored to it. However, the real cost-saving benefits of mass timber can only be realized if time and resources are invested in thorough pre-planning. The higher initial material costs can be offset by the scheduling and erection efficiencies, making mass timber a financially viable option in the long run. Constructors are responsible for improving in the areas of planning, scheduling, and communication to ensure the successful execution of projects.

2.6 CHAPTER QUESTIONS

1. Major capital projects have been identified as being behind schedule in ___% of cases?
   a. 10
   b. 15
   c. 20
   d. 80
2. Not only do mass timber structures require less labor, that labor is able to outproduce traditional cast-in-place structures?
   a. True
   b. False
3. What is a main consideration project teams must build into their schedule, when working with a carpentry team lacking experience installing Mass Timber?
   a. Learning curves
   b. An overall decrease in productivity
   c. Bigger crew sizes
   d. None of the above
4. What are the main scheduling factors to consider on a Mass Timber project? (Choose all that apply)
   a. Learning curves
   b. Smaller crew sizes
   c. Increased productivity
   d. Adjusted critical path
   e. Subsequent subcontractor coordination
   f. All of the above
5. Due to mass timber's relative infancy in the United States market, teams should be aware of which of the following shifting trends when developing schedules (choose multiple).
   a. Engineering efficiency
   b. Supply chain maturation
   c. Increase in builder's expertise

   d. Increase in shipping time
   e. Manufacturing Efficiency
   f. All of the above
6. Which of the following factors of mass timber construction contribute to scheduling efficiencies?
   a. MEPFP manufactured penetrations
   b. Reduced need for shoring
   c. No curing times
   d. Smaller crew sizes
   e. **All of the above**
   f. A and B
7. Data show that mass timber structures are approximately 15–30% faster to erect when compared to traditional steel and concrete structures.
   a. **True**
   b. False
8. The growth of which of the following is predicted to have a major impact on the efficiency of mass timber:
   a. Artificial intelligence
   b. Builder expertise
   c. Maturation of the supply chain
   d. Prefabrication technologies
   e. **All of the above**
   f. None of the above
9. Mass timber projects require smaller concrete foundation work due to the lighter weight of mass timber compared to steel and concrete structures.
   a. **True**
   b. False
10. Small things add up; which of the following choices are referenced in the chapter that has a direct, positive effect on the schedule of mass timber projects.
    a. Adhesives
    b. **Fasteners**
    c. Weather protection
11. Mass timber projects are often aligned with which of the following methodologies that eludes to the efficiency of delivery and installation of mass timber materials. This often helps alleviate the need for storage and limits damage potential.
    a. **JIT**
    b. Off-site storage
    c. On-site storage
    d. All of the above
12. Subcontractor sequencing on mass timber projects must be approached the same way as structural steel and cast in place concrete structures.
    a. True
    b. **False**

13. ______ times vary depending on numerous factors; including but not limited to manufacturer location, global supply chain variables, design, and workforce availability.
    a. Installation
    b. Payment
    c. Storage
    d. **Lead**
14. Mass timber projects can actually be erected too quickly, with examples showing that mass timber projects have been erected so quickly that subsequent contractors cannot keep up.
    a. **True**
    b. False
15. Significant cost and time savings of published studies need to be shared with subcontractors in order to increase accuracy of estimated and schedules alike.
    a. **True**
    b. False

REFERENCES

Agarwal, R., Chandrasekaran, S., & Sridhar, M. (2016, June 24). Imagining construction's digital future. McKinsey & Company. www.mckinsey.com/capabilities/operations/our-insights/imagining-constructions-digital-future

Beck, D. (2023, January 27). How to set up your next mass timber construction project for Success. Building Design+Construction. www.bdcnetwork.com/home/news/55154404/how-to-set-up-your-next-mass-timber-construction-project-for-success

Corgan. (2023). Designing with mass timber. Comparative analysis of mass timber and concrete prototypes. www.corgan.com/sites/default/files/inline-files/Designing%20with%20Mass%20Timber.pdf

Mirando, A., & Onsarigo, L. (2022). Construction productivity comparison between cast-in-place concrete and mass-timber framing: A case study of the nation's largest mass-timber building. In *ASC2022 58th Annual Associated Schools of Construction International Conference* (Vol. 3, pp. 443–451).

Onsarigo, L., & Mirando, A. M. (2023). Quantifying MEPFP Mass-Timber Trade Efficiency through Vertical Mechanical Fastener Analysis. *Mass Timber Construction Journal*, 6(1), 17–23.

WoodWorks. (2022, January 21). Wood Product Council. Holes and penetrations in mass timber floor and roof panels. www.woodworks.org/resources/holes-and-penetrations-in-mass-timber-floor-and-roof-panels/

3 Estimating Mass Timber Projects

"To estimate effectively is to anticipate the unexpected"—**Author Unknown**

LEARNING OBJECTIVES

After studying this chapter, the reader should be able to:

- Understand the cost estimating process
- Grasp cost concepts specific to mass timber
- Identify hard and soft cost considerations specific to mass timber
- Recognize current and future cost trends in mass timber construction
- Understand the construction manager's role in estimating mass timber projects.

3.1 INTRODUCTION TO CONSTRUCTION ESTIMATING

While cost is just one of the five pillars of construction, it often serves as the primary driver of decision-making. Accurately determining construction costs has always been a challenge, but 21st-century factors such as globalization, evolving delivery methods, pandemics, commodity fluctuations, and other external variables have added new layers of complexity. This chapter explores these emerging cost considerations in the commercial construction field, with a particular focus on mass timber structures. Estimating costs for mass timber projects differs significantly from traditional construction, and understanding the unique nuances of mass timber is crucial for making accurate cost projections.

Construction teams often collaborate closely with developers who must decide between traditional structural systems and mass timber. During this phase, cost estimates are typically fluid, making it essential for the construction team to provide precise and informed input. In addition to construction costs, developers must also consider several other costs that play a critical role in the overall financial feasibility of a mass timber project including financing, design, maintenance,

DOI: 10.1201/9781003482543-3

insurance costs, etc. The decision to proceed with a mass timber project typically rests with the development and financial teams rather than the construction team. As we've discussed, successful mass timber projects depend on early collaboration between development and construction teams in the pre-construction phase. The construction team's ability to influence the "go/no-go" decision is critical, as the unique delivery methods of mass timber offer construction teams additional opportunities to enhance project value and expand services.

The goal of an estimator is to achieve the highest possible accuracy in determining costs and quantities. To do so, they must possess specialized knowledge across multiple disciplines. Estimators need to understand their clients' needs, assess project risks, and analyze company data, pricing, and global market trends. As construction managers are increasingly tasked with broader responsibilities, estimating continues to be a key focus. Estimators should leverage every available tool and resource to provide accurate cost projections for mass timber projects.

This section provides significant value to readers by offering real-world mass timber cost data and discussing the challenges teams are currently facing. Hidden costs associated with mass timber, which are often overlooked, can be a decisive factor in choosing a structural system. Here, we highlight some of these lesser known, but critical costs tied to the mass timber delivery and construction model.

3.2 THE ESTIMATING PROCESS

The estimating process involves analyzing the work to be performed, which includes assessing the nature of the work, the quantity of work, required resources (such as equipment and labor), their respective production rates, and specific jobsite conditions that may impact project costs. Different types of estimates are developed throughout the design and construction phases. Initial estimates, based on limited information, tend to be less accurate. As the design progresses and more information becomes available, subsequent estimates become increasingly precise. However, with mass timber, precise estimates can be achieved earlier in the design phase due to the integrated delivery approach.

There are two primary types of estimates that project teams typically develop and that are discussed in this section.

3.2.1 Conceptual Estimates

Conceptual estimates, also known as preliminary, approximate, or budget estimates, are among the first attempts to predict project costs before a detailed and complete design is available. Other terms commonly used to refer to conceptual estimates include feasibility estimates, early estimates, authorization estimates, equipment-factored estimates, order-of-magnitude estimates, top-down estimates, screening estimates, or rough estimates. When prepared by or for the owner, conceptual estimates are typically part of the feasibility study, helping the owner decide whether to proceed with the project.

During various phases of architectural design—such as schematic design, design development, and construction documents—conceptual estimates are developed by the designer to give the owner an idea of what the project may cost and to inform design decisions that impact project costs. Conceptual estimates prepared by the designer can vary in accuracy based on the available information:

- **Level I** is prepared with little to no design and is accurate within +40% and –10%.
- **Level II** is prepared after completing the preliminary design and is accurate within +25% and –5%.
- **Level III** is prepared once the final design is complete and is accurate within +10% and –3% (Peurifoy & Oberlender, 2013).

Conceptual estimates are also prepared by contractors to assist with "go/no-go" decisions during the bidding phase. Before committing resources to prepare a bid package, contractors must assess whether it is suitable to bid on the project. Some of the primary variables in this decision include available bonding capacity, skills and resources, working capital, the nature of the project, and its location. Approximate estimates help determine whether sufficient working capital and bonding capacity are available to take on the project.

Making Conceptual Estimates

Conceptual estimates are based on experience with similar projects and historical data. Past projects, similar to the one being considered, are invaluable for generating reliable estimates. Adjustments can be made to the reference project data to better predict the cost of the current project. Several methods can be used to make conceptual estimates, which are briefly described here. For more detailed information, an estimating text can be consulted.

Conceptual Cost Estimating Methods

Unit Estimating: This method involves relating the total cost of a project to specific building units that significantly influence or predict the overall cost. By using these units, estimators can draw on historical data from similar projects to make informed cost projections for a new project. For example, in a school construction project, the number of students can serve as a unit. A unit rate, such as the cost per student derived from historical data on similar school projects, can be used to estimate the cost of a new school based on its projected capacity.

This method is widely applicable in various types of construction projects, where cost factors can be linked to measurable and predictable building units. Common examples of units used for cost estimation include:

- Cost per square foot of floor area for commercial or residential buildings
- Cost per bed in a hospital or healthcare facility
- Cost per room in a hotel or motel
- Cost per seat in venues such as stadiums, theaters, or restaurants

Unit estimating provides a quick and effective way to generate cost projections for different types of projects, especially during the early phases of design or when working with limited information. By leveraging historical data and known unit rates, estimators can deliver more accurate cost predictions based on well-established patterns and benchmarks.

Broad-Scope Conceptual Estimating: This approach uses cost data from multiple completed projects that are similar to the proposed project. The method calculates a weighted unit cost (UC) by taking into account various historical cost data points to generate a more accurate forecast. The formula used for this calculation is:

$$\mathbf{UC = ((A + 4B + C))/6}$$

where:

UC = Forecast unit cost
A = Minimum unit cost from past projects
B = Average unit cost from past projects
C = Maximum unit cost from past projects

This formula provides a balanced approach by giving more weight to the average cost data (B), which tends to represent the most typical scenario. The resulting forecasted unit cost is then adjusted to account for various factors that may affect the current project, such as:

1. **Location**: Adjusting for location indices to reflect regional cost variations (e.g., labor, material, and equipment costs may differ by region).
2. **Time**: Adjusting for inflation or the time value of money to ensure that the estimate reflects current market conditions.
3. **Size**: Adjusting for the scale or size differences between the historical projects used for comparison and the proposed project.

Broad-scope conceptual estimating is especially useful during the early design or feasibility stages of a project, where detailed cost data may not yet be available but a rough estimate is needed to inform decision-making. By leveraging historical data from similar projects and adjusting for key variables, this approach provides a solid basis for initial cost forecasting.

Narrow-Scope Conceptual Estimating: This method is similar to broad-scope estimating but focuses on more specific, project-related parameters. In narrow-scope estimating, the project is first broken down into key components or materials that are critical to the project's construction, for example, cubic feet of concrete, tons of steel, board feet of dimension lumber, and linear feet of piping.

Once these parameters are identified, the estimate is developed using historical cost data associated with each specific material or component. The historical data are then adjusted for location, time, and size as in broad-scope conceptual estimating method.

Narrow-scope conceptual estimating is particularly useful when detailed information about certain project components (e.g., specific materials or systems) is available early on, but the full design or scope is still being developed. It provides a more targeted estimate than broad-scope estimating, allowing project teams to better understand the costs associated with specific parts of the project while still maintaining flexibility for later adjustments.

Other Conceptual Estimating Methods

Other methods include:

Parametric Cost Estimate: This method utilizes statistical analysis of historical data to develop models that predict costs for proposed projects. It considers various parameters such as location, size, materials, complexity, and other relevant cost data. By analyzing these variables, parametric estimating provides a more accurate and data-driven forecast for project costs.

Cost Capacity Curves: These are graphical representations that account for economies of scale, helping to provide more accurate estimates for projects that are either upscaled or downscaled. By analyzing how cost changes with project size, cost capacity curves offer valuable insights into the financial impact of scaling a project up or down, ensuring a more precise cost estimate.

3.2.2 Detailed Estimates

Detailed estimates are prepared during the bidding phase, after the design is complete and most information is available. These estimates are typically done by contractors submitting bid proposals, hence the term "bid estimate." Proper estimating involves analyzing the work to be done, including the type and quantity of work, the required equipment, crew production rates, materials needed (permanent and construction), and other project or jobsite conditions that may affect the schedule and cost. Detailed estimating is a multi-step process that includes:

- Reviewing the bidding documents
- Conducting a site visit to assess the conditions
- Performing a material quantity takeoff
- Obtaining costs from vendors and suppliers
- Obtaining subcontractor quotes
- Determining indirect costs (insurance, bonds, permits, fees, taxes, overhead)
- Determining contingency and profit

There are generally seven major elements in a detailed project cost estimate: material cost, labor cost, equipment cost, subcontractor costs, overhead, contingency, and profit.

Material Cost

Estimating material costs in construction is generally a straightforward process, but it requires precision to ensure accuracy. A quantity takeoff from the construction drawings is used to determine the exact quantity of materials required for the project. This process involves identifying all the materials necessary, based on detailed measurements and specifications provided in the drawings.

Advancements in technology have significantly improved the efficiency and accuracy of material takeoffs. Today, tools and software are available that can automatically extract material quantities from digital construction prints or 3D models. These tools can streamline the process, reduce human error, and save valuable time for contractors and estimators.

Once the quantities are determined, the contractor solicits quotes from various vendors and suppliers. These quotes are typically time-sensitive, meaning they are valid only for a limited period. Therefore, it is crucial for the estimator to ensure that the quotes remain valid long enough to cover the duration needed for awarding the contract and placing the order. This often involves managing timelines carefully and maintaining communication with suppliers to avoid discrepancies or price fluctuations during the procurement phase.

Labor and Equipment Cost

Labor and equipment costs are fundamental components that influence every phase of a construction project, from assembling components and fabricating pieces to operating heavy machinery. These costs are primarily driven by two key factors: cost per unit time (typically calculated on an hourly basis) and productivity rates.

The cost per unit time reflects the hourly wage or rental cost associated with labor and equipment. These rates can vary based on the type of work being performed, the skill level of the workers, and the type of equipment being used.

Productivity rates, on the other hand, measure how efficiently labor and equipment complete a specific task within a given time frame. These rates can be influenced by a range of factors, including:

- **Crew Composition**: The number and skill levels of workers involved.
- **Individual Performance**: Variations in the efficiency of each worker.
- **Project Conditions**: Site-specific factors like weather, terrain, or worksite accessibility.
- **Experience and Skill**: More experienced or highly skilled workers generally have higher productivity rates.
- **Supervision**: Effective management and oversight can help optimize workflow and minimize delays.

When estimating labor and equipment costs, historical data from past projects can be a valuable source for determining productivity rates. If historical data are not available, estimators can turn to published cost manuals or industry benchmarks, which provide general guidelines for typical productivity rates based on task type and work conditions. By incorporating accurate productivity rates into cost calculations, estimators can more effectively forecast labor and equipment expenses, helping to create a more reliable and realistic project budget.

Subcontractor Costs

Subcontractor costs are typically determined by obtaining quotes from the subcontractors who will perform specific portions of the work. When the prime contractor is assembling a bid package, they generally follow these steps to manage subcontractor costs:

- **Breakdown of the Project Scope:** The first step is to break the project into distinct portions—those to be self-performed by the prime contractor and those to be subcontracted out. The goal is to ensure that the entire project scope is addressed, with clear division of responsibilities and no overlap between the prime contractor's and subcontractors' tasks.
- **Soliciting Subcontractor Quotes:** The prime contractor requests multiple quotes from subcontractors for each portion of the work being subcontracted. It is important that these quotes include all necessary components:
  - <u>Material Costs:</u> Breakdown of materials needed for the job.
  - <u>Labor and Equipment Costs:</u> Details of labor rates, hours, and equipment usage costs.
  - <u>Activity Durations:</u> The anticipated length of the task(s) within the assigned scope.
  - <u>Subcontractor Overhead:</u> Costs associated with running the subcontractor's business.
  - <u>Profit:</u> The subcontractor's markup or profit margin.
- **Assessing Subcontractor Qualifications:** The prime contractor evaluates the subcontractor's qualifications to ensure they have the capability to complete the work. This includes reviewing:
  - <u>Experience:</u> The subcontractor's relevant experience with similar projects.
  - <u>Financial Stability:</u> Ensuring the subcontractor has the financial resources to meet project demands.
  - <u>Available Equipment:</u> Confirming that the subcontractor has the necessary equipment to complete the work.
  - <u>Licensing, Insurance, and Bonding:</u> Ensuring that the subcontractor has the appropriate licensing, insurance, and bonding required for the project.
  - <u>Reliability:</u> Assessing the subcontractor's track record for meeting deadlines and delivering quality work.
- **Adding Contingency for Subcontractor Work:** A contingency is added to the subcontractor's portion of the work to account for unforeseen issues that may arise during project execution. This helps cover potential risks, such as delays, material shortages, or unanticipated site conditions.

Typically, a percentage (usually 5–10%) is added to manage these uncertainties.

- **Adding Overhead and Profit (O&P):** The prime contractor adds their own Overhead and Profit (O&P) on the subcontracted work. This is primarily to cover administrative costs associated with managing subcontractors and overseeing the project. O&P ensures that the prime contractor can fund project management, coordination, and other essential administrative functions.

By following these steps, the prime contractor ensures that subcontractor costs are comprehensive, competitive, and well-managed, ultimately contributing to the successful execution of the project.

Contingency

A contingency is a portion of the project budget reserved to cover unforeseen costs or risks that may arise during the construction process. It acts as a financial buffer, ensuring that the project can continue smoothly in the event of unexpected changes, challenges, or events. In construction, changes are common due to the dynamic nature of the industry, which is why having a contingency plan in place is crucial for mitigating the impact of these uncertainties.

Contingency is typically calculated as a percentage of the total project cost, with the percentage generally ranging from 5% to 20%. The exact amount depends on the project's risk level. High-risk projects—such as those with complex designs, incomplete or poorly detailed plans, unknown site conditions, tight schedules, or unproven technologies and novel materials—are more likely to require a larger contingency.

Overhead and Profit

Overhead refers to the indirect costs required to complete a project but are not directly tied to construction activities. These costs are essential for the business's operation and include expenses such as project management staff, office rent, salaries and benefits, depreciation, marketing, advertising, and insurance. Essentially, overhead covers the fundamental administrative and operational costs of running a construction company.

Profit in construction management is inherently subjective, with various factors influencing the development of profit margin goals. By definition, profit in construction is the amount of money a company has left over after all project costs (both direct and indirect) have been subtracted from revenue. Profit margins in construction projects are highly variable and cannot be generalized. They depend on multiple factors, including the scale of the project, the type of work, and the complexity of the design and construction processes.
The basic formula for project profit is:

$$\text{Project Profit} = (\text{Project Revenue}) - (\text{Overhead} + \text{Direct Costs})$$

Profit goals for site-specific projects are calculated based on several factors, including current economic conditions, the relationship with the project owner, site-specific risks, organizational goals, design risk, schedule timelines, contractual stipulations, the company's overhead profile, project size, supply and demand, governmental policies, and the organizational structure of the construction company.

Generalized profit goals for construction management companies typically range from **5% to 10%**, with specialty subcontractors often aiming for higher profit percentages due to the complexity and specialization of their work.

3.3 MASS TIMBER COSTS

Significant price variations occur throughout the estimating process for mass timber projects. This section outlines the primary cost differentials associated with these projects. Domestically, mass timber materials come with an upfront premium of approximately 8–12% when compared to traditional structural steel and concrete. Estimating teams must also factor in fluctuations in raw lumber prices. In addition, soft cost premiums are currently common, which will be explored further in this section. However, these initial cost premiums can be offset by benefits such as smaller crew sizes, faster and more efficient erection processes, improved productivity in subsequent trades, and higher rental or sales premiums. It is essential for estimating teams to integrate these cost differentials into their estimates to ensure a comprehensive and accurate cost analysis.

3.3.1 Hard and Soft Costs

Current challenges in commercial mass timber projects often stem from premiums in soft costs. Soft costs refer to project expenses that are necessary for completion but not directly tied to physical construction. Commonly referred to as overhead, these costs often include architectural and design fees, legal and accounting expenses, loan interest, project management fees, sustainability assessments, taxes, advertising and public relations, and construction insurance. Unlike hard costs, soft costs are typically managed separately and distributed over the entire project timeline, making them more difficult to predict and estimate.

In contrast, hard costs are directly related to physical construction and generally cover materials, labor, and equipment. Examples of hard costs include mass timber premium costs, water mitigation costs, drywall costs, and MEPFP (mechanical, electrical, plumbing, fire protection) costs. It is critical for project teams to understand the cost differentials between soft and hard costs in mass timber projects.

3.3.2 Material Premium

Currently, mass timber material premiums exist both domestically and internationally when compared to traditional building materials. As of 2024, it remains more cost-effective for project teams to source mass timber from European and Canadian suppliers rather than domestic manufacturers. Domestic premiums over steel and concrete are approximately 10%, while international premiums are around 7%. The higher price for domestic products can be attributed to several factors, including the nascent state of domestic supply chains, the absence of established manufacturing processes, and challenges related to raw material acquisition, particularly due to stringent harvesting guidelines.

Major domestic manufacturers are only beginning to produce mass timber at scale and are working to optimize their distribution channels. Significant domestic investment is required to compete with European and Canadian manufacturers, who have been producing mass timber for over two decades. As discussed in the manufacturing section, the high level of technological integration by major international manufacturers allows for greater efficiency giving them a competitive edge over the domestic market.

It is predicted that domestic cost premiums will decrease as mass timber becomes more established in the market. However, achieving this will require a broad and multifaceted shift in the factors affecting cost. Suppliers will need access to more raw timber, a challenge given the constraints of regulations and laws. Significant investment is also necessary to remain competitive with long-established manufacturers, and securing the capital required for this transformation will take time. While upfront structural material premiums must be factored into estimates, which often deters project teams, these premiums can be offset by experienced construction teams who implement strategies to maximize the efficiency of mass timber's properties during the construction phase.

3.3.3 Drywall Cost

Drywall is a building material used in nearly every project, and while it may seem like an unusual cost element to consider due to its standardized nature, its use on mass timber projects varies significantly from project to project. This variability stems from its close ties to fire protection, design, and building classification. Mass timber is often celebrated for its aesthetic appeal, which encourages design teams to maximize the exposure of the timber. However, municipalities frequently debate the amount of drywall coverage required for fire mitigation.

For example, in the project discussed in the water mitigation section, negotiations over drywall coverage were ongoing even as drywall was being installed. Initially, the municipality required 75% of the interior structure to be covered in drywall. After extensive discussions and presenting data on mass timber's fire-resistant properties, the development team successfully reduced this requirement to 51%. In projects of this scale, even small adjustments in drywall coverage can have significant implications for aesthetics, cost, and project timelines.

Project teams often negotiate drywall coverage requirements even after the construction drawings have been approved. Therefore, it is recommended that drywall coverage be finalized during the pre-construction phase, prior to the start of construction, to ensure proper planning and accurate cost estimates. Figure 3.1 illustrates the relationship between drywall and mass timber exposure, with teams needing to navigate coverage requirements based on fire codes, municipal regulations, and interior design considerations.

FIGURE 3.1 Drywall Cover versus Exposed Mass Timber.

3.3.4 Mechanical Electrical Plumbing and Fire Protection Costs

Contractors often increase bid costs when working with unfamiliar materials or new construction methods, which has contributed to higher subcontractor bids on mass timber projects. However, recent studies have provided data that can help address these concerns. As contractors gain more experience with mass timber, bids should begin to reflect that familiarity, leading to a decrease in premiums.

One study, for example, highlighted the efficiencies achieved by MEPFP contractors when installing overhead conduits, fixtures, ductwork, and other critical systems in wood structures (Figure 3.2). The study found that installing fasteners into wood, rather than concrete, resulted in significant time savings. On a 512,000-square-foot project, these efficiencies saved over 595 man-hours, translating into a straight-time cost savings of more than $42,000. Furthermore, the reduction in the number of bits and drills required to predrill fasteners resulted in additional savings of $53,400 for the MEPFP trades (Onsarigo & Mirando, 2023).

Data presented to subcontractors during the bidding process for mass timber projects can play a crucial role in helping them understand the material's potential for improving efficiency. As more studies are conducted and contractors gain greater familiarity with mass timber, both bid and installation costs are expected to decrease. To foster competitive bidding, it is essential that this data be shared with subcontractors upfront. By providing insights into the efficiencies and cost savings mass timber can offer, subcontractors will be better equipped to accurately assess the project details, resulting in more informed and competitive bids.

FIGURE 3.2 A Tradesman Installing Overhead MEPFP Systems.

3.3.5 Insurance

Insurance is a significant soft cost in commercial construction projects and can often be the deciding factor in whether a building will be constructed. There are two key types of insurance that project teams must consider. The first, and more crucial during the construction phase, is the builder's risk policy. This policy protects teams from risks associated with the construction process. The second, which is more important to the ownership team, is the post-construction fixed property insurance. This policy covers the building during the occupancy phase, protecting against risks such as moisture, flooding, fire, and wind damage.

During the estimating process, the construction team typically includes the builder's risk policy in its overall project costs. Meanwhile, the ownership team usually accounts for the cost of fixed property insurance in their own cost analysis. Builders risk policy costs are determined based on several key factors:

- **Overall Value of the Project**: including cost of materials, equipment, labor, soft costs, and contingency.
- **Construction Duration**: the longer the project timeline, the more risk, resulting in higher premiums.
- **Location and Risk Factors**: geographical location, climate, susceptibility to natural disasters, and other location-specific factors will affect premiums.
- **Type of Construction**: unique designs and complex projects with higher risk such as historical renovations and large-scale commercial developments, can increase premiums.
- **Deductible and Coverage Limits**: higher deductibles generally lead to lower premiums, while higher coverage limits typically result in higher premiums. It is crucial for project teams to find a balance between deductibles and

coverage limits to optimize costs. Balancing these factors ensures that the project remains financially secure while managing overall risk effectively.

- **Relationship and Track Record**: construction teams are often assessed based on their risk profile when it comes to insurance rates. Major companies with a strong track record of delivering projects with minimal claims secure better rates, as their low-risk history makes them more attractive to insurers. Additionally, construction teams that have long-standing relationships with insurers are likely to benefit from more favorable rates due to the trust and consistency built over time. These factors play a key role in helping construction teams lower their insurance costs while maintaining adequate coverage for their projects.

The factors outlined above make it difficult to establish a universal percentage for every project. However, insurance premiums typically range from 1% to 4% of the completed value of the structure, including construction costs. This range can have a significant impact on estimates, particularly for high-value commercial projects. As a result, it is crucial for project teams to negotiate effectively and gain a clear understanding of these costs through the insurers they work with. By carefully evaluating various insurance options and considering their effects on the overall project budget, teams can make more informed decisions that help manage costs without sacrificing coverage.

For mass timber projects, teams should expect to be on the higher end of this range, as the current insurance market views mass timber as a unique building material. Insurers are often cautious with new building methods, and mass timber is no exception. Although mass timber may represent only a portion of the overall project, insurers remain risk-averse, and to address this, they often raise premiums toward the upper end of the range. In some cases, insurers may even decline to cover mass timber projects due to their distinct nature.

Fortunately, resources are available to help project teams navigate the complexities of insuring mass timber structures. Estimating cost premiums can be a fluid and challenging process. To aid in this, WoodWorks published the *Mass Timber Insurance Playbook*, which provides valuable insights for project stakeholders involved in the insurance and risk management process. It is recommended that project teams take full advantage of these resources to inform their decision-making regarding insurance.

3.3.6 Escalation

Escalation in construction refers to the increase in costs for labor, materials, or other project expenses over time. Factors such as changes in project scope, materials, design, and building specifications can all affect the cost of estimated line items throughout the project, particularly for materials tied to commodity price fluctuations. For instance, supply chain disruptions can drive commodity prices higher. The COVID-19 pandemic underscored the volatility of global supply

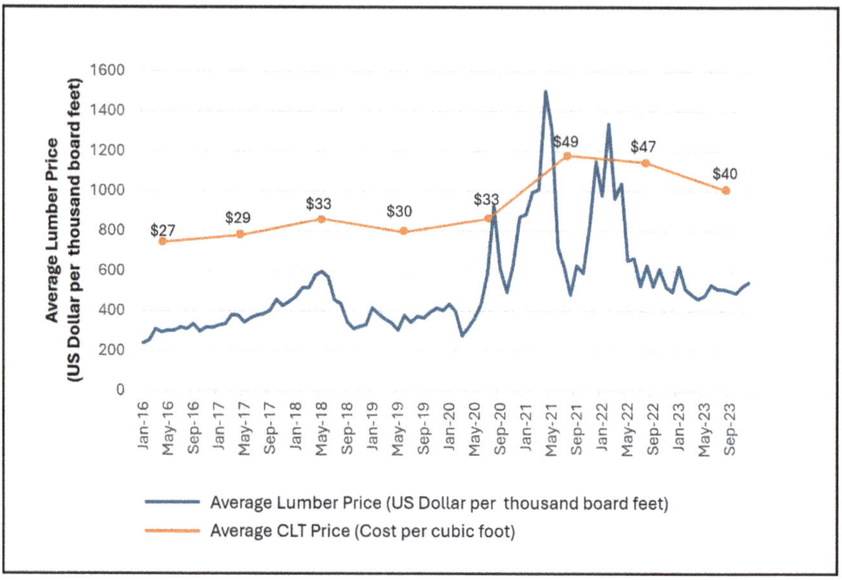

FIGURE 3.3 CLT's Market Cost Relationship to Raw Lumber Pricing Fluctuations.

chains and tested project contracts and escalation clauses. Prior to the pandemic, suppliers and contractors were often able to lock in product prices for extended periods. However, in today's commercial construction contracts, material quotes generally have shorter lifespans and can change rapidly in response to shifts in global commodity markets. Mass timber products are not immune to these price fluctuations, as changes in raw lumber prices can directly affect the quotes.

Manufacturers of mass timber products do have some protection, as many purchase raw lumber in bulk and can secure pricing for longer periods compared to other suppliers. However, project teams should still factor in escalation price projections before finalizing contracts with manufacturers. EWP pricing will fluctuate based on raw lumber prices and should be reflected in project estimates. The *Mass Timber Price Index (2024)* provides data that compare lumber prices with the cost of producing CLT panels (Figure 3.3). It is estimated that 40–60% of CLT manufacturing costs are driven by raw lumber prices. Therefore, project teams should anticipate these price fluctuations and ensure that quotes for mass timber production are secured from EWP manufacturers.

3.3.7 Contingencies

Construction contingencies are funds set aside in the project's estimate to cover unexpected costs. As a risk management tool, these predetermined funds help ensure that a project can be completed on time and within budget. Contingencies are typically used to address issues such as design errors, weather delays, supply chain disruptions, scope changes, rework, and personnel costs. As project risks and uncertainties increase, teams should consider raising the contingency amount. Over time, as teams gain more experience with mass timber projects, they may be able to reduce contingencies if the unknowns primarily involve mass timber.

There is no fixed dollar amount for construction contingencies; they are usually expressed as a percentage of the total project cost. Project teams can typically allocate up to 10% of the budget to the contingency fund following an assessment of the potential risks. The amount is typically negotiated early with the client and depends on various factors. The type of project plays a key role in determining the appropriate percentage. For example, major renovations, historical renovations, and unique building designs generally require a higher contingency percentage to account for increased project risks and unknown conditions. If the construction team lacks previous mass timber experience, it is recommended to use the higher end of the contingency percentage range.

3.3.8 Storage

An offsite storage facility may be necessary for project teams, depending on various factors. If the team determines that a storage site is needed for the duration of the mass timber erection, the associated costs should be included in the project estimate. Some contractors may have their own storage facilities, while others may need to rent one. The cost of offsite storage can vary widely depending on factors such as geographic location, facility size, and the type of mass timber products being stored. Additional variables, including storage duration, facility type, and the volume of materials, can also affect the cost. Generally, storage fees range from $1 to $5 per square foot per month.

Mass timber components are often shipped without pre-attached connections to enhance shipping efficiency. These connections are usually manufactured by third-party fabricators and delivered separately to the job site or a storage facility. Since mass timber components are typically transported in containers or on flatbed trucks or trains, protruding connections can hinder stacking and storage. To address this, many teams opt to install the connections in a warehouse before transporting the completed pieces to the job site for installation (Figure 3.4). Beyond connection installation, teams should also account for weather protection, specialized hoisting equipment for handling mass timber materials, and security measures at the storage facility.

3.3.9 Water Mitigation Cost Considerations

As discussed in the quality chapter, every mass timber project requires a water management plan, and each type of plan involves different financial considerations, which will be outlined here. The estimator needs to ensure that the costs related to water management are accounted for in the project estimate. Manufacturers offer a range of water mitigation options, each with its own costs. Project teams can choose to rely on the manufacturer's preventative measures, opt for no mitigation, manage

FIGURE 3.4 Beam Delivered with Connection Installed in an Off-site Warehouse Setting.

water protection themselves, or use a combination of these options. Each choice has its pros and cons, with cost being a key factor. Estimators should collaborate early with the manufacturing, design, and construction teams to determine the best option for their specific project.

3.3.9.1 Manufacturer-Driven Water Mitigation

Manufacturers offer several preventative measures to help protect mass timber components, with costs varying by option. It is advisable to discuss pricing and availability directly with manufacturers. Key manufacturer-driven mitigation strategies include:

- **Sealers at the Production Facility:** A variety of sealers, available at different price points, can be applied during production to enhance moisture resistance.
- **Vapor Wrap for Delivery:** For added protection against moisture, a protective wrap can be applied to timber during transit, typically at an additional cost.
- **Just-in-Time Delivery:** Some manufacturers provide storage at their facility and deliver materials precisely when needed for installation,

reducing prolonged exposure to the elements and minimizing on-site storage requirements.

- **On-Site Unloading:** Certain manufacturers allow project teams to keep materials secured on the flatbed trailer upon arrival, reducing handling and exposure during the unloading process.

3.3.9.2 Construction Phase Water Mitigation:

- **Do Nothing:** Project teams may choose not to implement protective measures, accepting potential remediation costs and schedule impacts during project closeout. Those opting for this approach should budget accordingly for potential repairs and aesthetic remediation. Remediation costs will vary based on the extent of exposure-related damage, and teams are advised to collaborate with subcontractors to develop more accurate cost estimates.
- **Comprehensive On-Site Water Mitigation Plans:** For teams actively managing water on-site, strategies vary based on project needs. Key measures include removing standing water (Figure 3.5), installing temporary and permanent roofing and drainage systems, and ensuring continuous timber protection. While exact costs depend on project specifics, teams should allocate dedicated labor and materials for water management efforts.
- **Storage and Just-in-Time Delivery:** Coordinating delivery and installation schedules on-site requires an upfront investment in planning, coordination, and trade alignment. This ensures that the site is properly prepared for both the timely delivery and efficient installation of materials. To shield mass timber products from weather exposure, teams can utilize protective storage solutions such as conex boxes, offsite warehouses, or temporary tarp structures.

3.4 OTHER CONSIDERATIONS IN ESTIMATING MASS TIMBER PROJECTS

3.4.1 Early Stakeholder Integration

As previously mentioned, the successful execution of a mass timber project hinges on the early integration of all relevant stakeholders. Delivery methods that promote early contractor involvement such as construction manager at risk, integrated project delivery, design-build, or design-assist are highly recommended. However, it's important to note that the costs associated with early integration and coordination must be factored into the owner's budget. In addition to the main contractor, certain first-tier subcontractors, particularly those involved in MEPFP systems, must also be brought in early during the design phase. These MEPFP engineers and subcontractors play a crucial role in designing (and modeling) their systems and ensuring their seamless integration with the overall project design. This proactive approach allows for the fabrication of mass timber components with the necessary system penetrations in place, ultimately streamlining construction and improving overall project efficiency.

FIGURE 3.5 Removing Standing Water Using Temporary Drain Piping.

3.4.2 Code Compliance and Permitting

Code adoption for mass timber is not consistent across the country, and the level of adoption can vary significantly by location. It is crucial to factor in the potential costs associated with obtaining additional approvals for mass timber projects, as local regulations may impose unique requirements. Local building codes, especially those related to mass timber such as fireproofing treatments and drywall requirements, can have considerable cost implications. For instance, many jurisdictions have not yet adopted the tall mass timber provisions outlined in the 2021 or 2024 International Building Code (IBC) (see Figure 3.6). This lack of uniform adoption can lead to delays, variability in design, and associated costs.

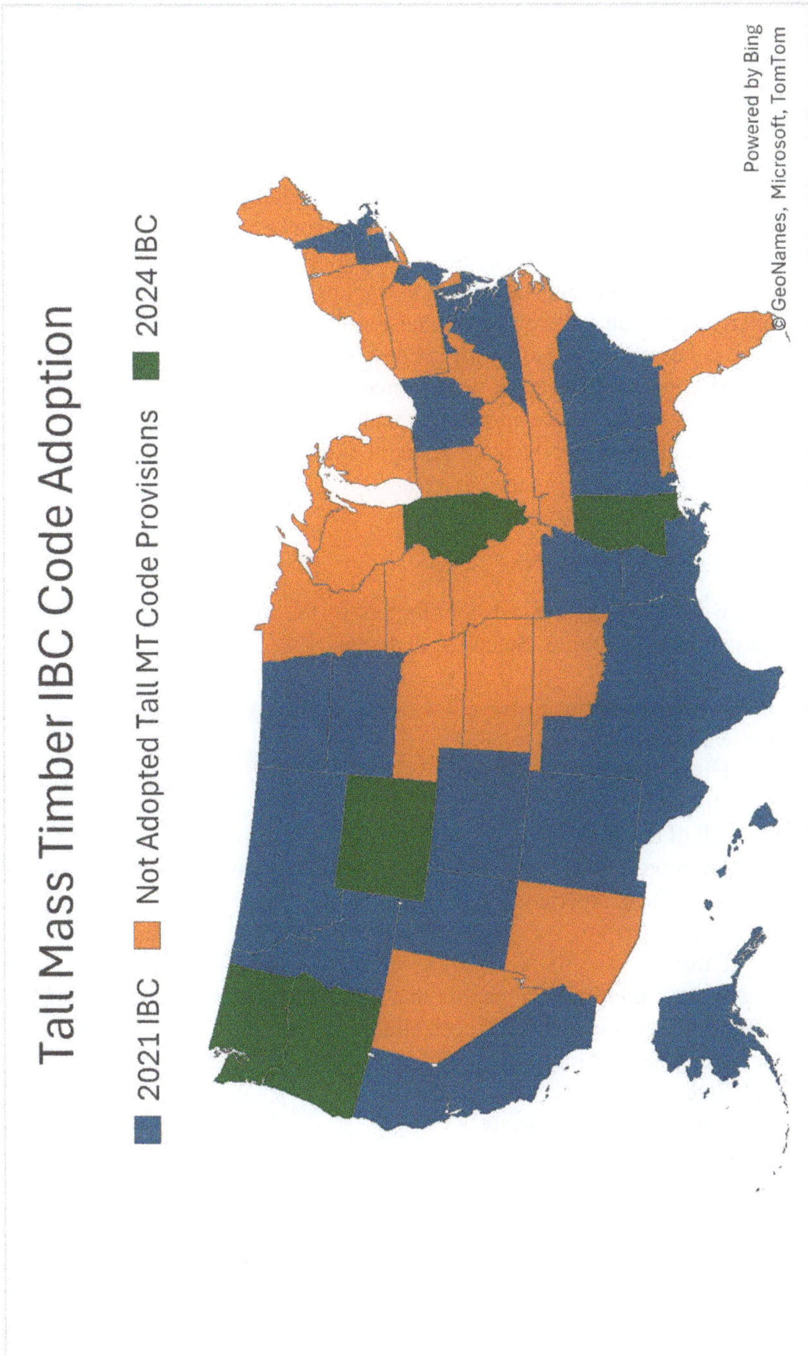

FIGURE 3.6 Mass Timber Code Adoption across the United States.

3.5 CHAPTER SUMMARY

To reduce errors and improve the accuracy of construction estimates, project teams must have a thorough understanding of the materials, processes, and variables specific to each scope of work. Owners are increasingly relying on construction teams to provide accurate estimates that directly impact project pro formas. Assembling the necessary knowledge to create these estimates can be challenging, which is why resources like this text and industry literature are invaluable for understanding the complexities of estimating mass timber projects.

Knowledgeable project team members play a crucial role in adding value to cost discussions. By understanding and accounting for the nuances discussed in this chapter, teams can ensure more precise cost projections. This chapter has covered key concepts in the estimating process, soft and hard cost considerations, subcontractor efficiency, technology for mass timber takeoffs, and essential resources for the reader. It is important to maintain open communication among all stakeholders, as costs can fluctuate throughout the development of mass timber projects.

3.6 CHAPTER QUESTIONS

1. There are three referenced contributors to the increase in rental rates in mass timber structures. Select all that apply:
   a. Visual appeal
   b. Lower operating costs
   c. Sustainability
   d. Higher profiles for renters
   e. Biophilic benefits
   f. All of the above
2. Construction contingencies should be increased on mass timber projects for teams that lack experience with using the material.
   a. True
   b. False
3. There are a multitude of factors that affect the cost of offsite storage for mass timber components. Choose all that apply:
   a. Geographical location
   b. Size
   c. Length of storage needs
   d. Machinery needed
   e. All of the above
4. Common mass timber codes stipulate that all projects are required to design the interiors with ____% of drywall coverage.
   a. 50
   b. 60
   c. 70

   d. There is no concrete number, it is subjective based on several factors
   e. None of the above
5. There are various levels of estimates. Which of the following levels represent an estimate prepared from little to no design provided and is accurate within +40% and −10% of costs?
   a. Level I
   b. Level II
   c. Level III
   d. Level V
6. The following are examples of hard costs in construction (choose multiple if applicable):
   a. Material
   b. Equipment
   c. Managerial costs
   d. Jobsite trailers
   e. Dumpsters
   f. Labor
7. There are a multitude of factors that contribute to the pricing structure of builders' risk policies. Select all factors that apply:
   a. Overall project value
   b. Duration of construction
   c. Site specific location and risk factors
   d. Type of construction
   e. Deductible limits
   f. Contractor track record of success
   g. **All of the above**
   h. A, B, D, and E
8. There are several areas of construction cost that will vary due to the utilization of mass timber components versus traditional methodologies. This includes which of the following? Choose all that apply:
   a. Builders risk rates
   b. Escalation factors
   c. Drywall cost
   d. Construction contingency
   e. **All of the above**
   f. None of the above
9. The relationship between CLT panels and raw material lumber pricing is estimated at what percentage range?
   a. **40–60%**
   b. 45–65%
   c. 15–20%
   d. 60–70%

10. What does the estimating process involve assessing
    a. Nature of the work
    b. Quantity of work
    c. Required labor
    d. Equipment production rates
    e. Specific jobsite conditions
    f. All of the above
11. Which of the following factors affect labor and equipment productivity rates?
    a. **Project conditions**
    b. **Experience and skill**
    c. **Level of supervision**
    d. Financial stability
    e. Insurance
12. What should a prime contractor consider when assessing a subcontractor's qualifications?
    a. Experience
    b. Financial stability
    c. Reliability
    d. Available equipment
    e. A and B
    f. **All the above**
13. ______________________refers to the increase in costs for labor, materials, or other project expenses over time in construction projects.
    a. Contingency
    b. **Escalation**
    c. Market fluctuations
    d. Price gouging
14. The successful execution of a mass timber project heavily relies on the integration of all relevant stakeholders after the design is sufficiently complete
    a. True
    b. **False**
15. Which of the following are manufacturer-driven mitigation strategies?
    a. On-site storage
    b. **Coating the MT with waterproof sealers**
    c. **Using a protective wrap during transit**
    d. Installing drainage systems

REFERENCES

Mass Timber Price Index (2024, November 20). International Mass Timber Conference. https://masstimberconference.com/report/content/mass-timber- price-index-2024/

Onsarigo, L., & Mirando, A. M. (2023). Quantifying MEPFP Mass-Timber Trade Efficiency Through Vertical Mechanical Fastener Analysis. *Mass Timber Construction Journal*, 6(1), 17–23.

Peurifoy, R., & Oberlender, G. (2013). *Estimating construction costs: Sixth Edition.* McGraw-Hill Higher Education.

4 Quality

"The bitterness of poor quality remains long after the sweetness of low price is forgotten."—**Benjamin Franklin**

LEARNING OBJECTIVES

After studying this chapter, the reader should be able to:

- Understand the QA/QC process.
- Identify the regulations and codes related to mass timber construction.
- Recognize the quality measures in the mass timber manufacturing process.
- Identify the damaging effects of moisture on mass timber products and effective water mitigation strategies.
- Understand different jointing systems and fasteners used in mass timber framing.

4.1 INTRODUCTION TO CONSTRUCTION QUALITY

Quality refers to how effectively a product or service meets its intended purpose while remaining free from defects. A high-quality product or project should not only be defect-free but also fulfill both explicit and implicit requirements as spelled out in the contract documents. In construction, ensuring quality means that the project: meets the needs of the owner or developer, functions as designed for its intended purpose, and complies with project specifications, industry standards, and regulatory codes.

Quality management is fundamental to the success of any mass timber project. Unlike other aspects of construction, where responsibility primarily falls on the construction team during execution, quality management is a shared responsibility among all stakeholders. The design, manufacturing, construction, and maintenance teams each play a crucial role in maintaining the structural integrity and aesthetics of the final product throughout the project's lifecycle. However, managing quality

DOI: 10.1201/9781003482543-4

during the construction phase presents unique challenges due to variable site conditions and external factors that can be difficult to control.

Quality management in mass timber construction involves a rigorous system of controls and inspections throughout design, manufacturing, and building processes. These measures ensure the structural integrity, durability, and safety of mass timber components through regular checks on moisture content, adhesive quality, and precise fabrication techniques. Industry standards and certifications, such as those from the American Wood Council and the Forest Stewardship Council, help uphold quality from the earliest stages of raw material acquisition.

Interestingly, many quality control (QC) practices in construction management have been adapted from manufacturing. While differences exist between the two fields, manufacturing QC methodologies can often be effectively applied in construction. Ensuring high-quality mass timber production is essential to the material's performance and longevity, which is a critical consideration for project developers.

This chapter outlines key timelines, checklists, and plans necessary to maintain QC across the manufacturing, delivery, installation, and post-construction phases of mass timber components. While various stakeholders contribute to ensuring product quality, the construction team ultimately bears the responsibility of meeting project specifications for the client. Achieving this requires active communication, proper training, thorough documentation, and adherence to best practices, all of which will be explored in this chapter.

4.2 QUALITY ASSURANCE AND QUALITY CONTROL (QA/QC) PROCESS

QA/QC measures are implemented to ensure that a construction project meets all necessary requirements for it to perform its intended function. While both QA and QC are essential in construction, they serve distinct purposes:

QA is a proactive, process-oriented approach focused on preventing defects before they occur. It involves establishing a systematic process to ensure that all construction activities adhere to predefined quality standards. QA includes defining project requirements, setting quality standards, providing training, and implementing procedures to maintain quality throughout the project lifecycle. Its primary goal is to ensure the overall process is robust, resulting in a final product that meets the desired quality standards.

QC, in contrast, is a reactive, product-oriented approach that focuses on identifying defects and deviations as the construction work progresses. QC involves monitoring, inspecting, and testing materials, evaluating workmanship, and conducting audits to ensure that the construction meets the specified requirements and regulations. When issues are identified, corrective actions are taken to address them.

Together, QA and QC ensure that construction projects fulfill their intended purpose and meet the required quality standards. QA emphasizes preventing issues, while QC focuses on detecting and resolving them.

4.2.1 Quality in the Manufacturing Process

Manufacturers uphold stringent quality standards when producing mass timber components, as their reputation depends on the quality of the final product. Construction teams must be well-versed in their manufacturer's QC requirements to ensure compliance with production standards as well as shipping and receiving protocols. Visiting manufacturing facilities offers valuable insights into these processes, and construction teams should be prepared to assess key factors during these visits.

QC in mass timber begins with sourcing raw materials. Manufacturers establish their own systems, methodologies, and manuals for production, often seeking third-party verification of their QC processes through certifications such as ISO 9000. Additional standards, such as ANSI A190.1, ensure adherence to strict quality criteria and remain the manufacturer's responsibility. These external certifications provide construction teams with confidence that the materials they purchase meet rigorous quality standards.

Manufacturers initiate their QC process with raw material acquisition. They ensure the timber is disease-free, mature, and meets production requirements. Upon arrival at the facility, each board is evaluated and tracked through each production stage to ensure traceability and quality assurance. At the start of the production line, the attributes of each piece of lumber are assessed, including moisture content, defects (such as warping, splitting, and decay), and their distribution. Through the use of technology, boards are sorted and grouped based on their characteristics and the production requirements for their end use. Controlling raw material quality is critical for manufacturers, as it directly impacts the final product's quality, efficiency, and ultimately saves time and money.

According to Nordic Structures (2020), QC systems at the manufacturing level should, at a minimum, include:

- Plant and process qualifications
- Testing of adhesive lots used for bonding timber components
- Daily QC operations, including:
    - Continuous, detailed checks at each stage of production, including in-line tests
    - Physical tests of representative samples
    - Visual inspections of finished products
- Regular audits conducted by accredited inspection agencies
- Production and review of plant manuals
- Production and auditing of QC records

Adhesives used to bond timber materials are of particular importance. Each new lot of adhesive should undergo qualification testing to assess its strength, wood failure rates, and durability before shipment. Construction teams should verify that manufacturers have obtained third-party quality verification credentials to ensure material reliability and performance.

Manufacturers typically use two key methods to assess their operations: process control and product checking.

1. **Process Control:** This method ensures consistent material production by monitoring the production process. It involves training and testing graders who visually inspect materials for quality. Additionally, machine calibration checks, consistency assessments, and other verification tools help maintain process accuracy.
2. **Product Checking**: This method evaluates the finished products to ensure they meet specification expectations. It identifies issues that may not be detected during process control, such as variations in raw materials.

4.2.2 General QA/QC on Construction Projects

An efficient and well-structured QA/QC process in construction projects helps reduce rework, improve quality, promote a safer work environment, and generate cost savings, all of which contribute to increased client satisfaction. A successful QA/QC process includes key components such as detailed drawings and specifications, submittal requirements, and a QC plan that outlines the procedures and responsibilities necessary to ensure quality. This plan encompasses quality processes such as inspections, testing, and quality reviews, all of which are further outlined in this section.

4.2.2.1 Project Specifications

Accurate and detailed drawings and specifications form the foundation of the QA/QC process. These documents serve as the baseline, providing comprehensive guidelines for contractors, subcontractors, and other stakeholders involved in the project, ensuring that all work is completed according to the required standards. Specifications play a critical role in defining the quality and performance standards for all aspects of the project, ensuring consistency and compliance across the board. There are several types of specifications, each serving a different purpose:

1. **Performance Specifications**: These specify the desired outcome or performance characteristics of a finished product or system, without detailing how to achieve them. They outline the functional requirements, such as load-bearing capacity or insulation value, but leave the contractor to determine the methods and means of construction to meet those outcomes.
2. **Prescriptive Specifications**: These specifications define both the quality of materials and the exact installation methods. They provide detailed instructions on how materials should be applied or assembled, including specific construction techniques, tolerances, and material standards.

3. **Proprietary Specifications**: These specify a particular product or brand, including the manufacturer and model name. Proprietary specifications limit the choice of materials or products to specific manufacturers, ensuring that only the exact product is used in the project.
4. **Reference Specifications**: These are specifications that reference established industry standards. They require the use of well-recognized, third-party standards or codes (e.g., ASTM, ISO), and ensure that the work complies with these pre-existing benchmarks without needing to detail the specific methods or materials.

By clearly outlining the materials, methods, and performance standards, these specifications help guide the construction process, ensuring the final product meets both design intent and regulatory requirements.

4.2.2.2 Submittal Requirements

A submittal is a written document and/or physical specimen provided by the contractor, manufacturer, supplier, or subcontractor to the owner's representative for approval before fabrication, delivery, and installation on the jobsite. Submittals are required by the design professional (engineer or architect) to ensure that the correct product is used in compliance with the drawings and specifications. There are various types of submittals, but the main ones include:

- **Shop Drawings**: Detailed drawings of prefabricated components or systems that show how they will be fabricated, assembled, and installed.
- **Product Samples**: Physical samples of products and finishes that will be used in the project, such as color swatches, countertop materials, and tile finishes.
- **Product Data**: Documents that communicate the key features and specifications of a product, including dimensions, characteristics, performance, and warranty.
- **Equipment Cut Sheets**: Documents detailing the specifications of equipment, including dimensions, capacities, model numbers, and performance data.
- **Mockups**: Physical models of building elements or components, used to visualize and test design features before actual construction begins.
- **As-Built Drawings**: Drawings that reflect the final product as completed or installed. These are updated to show any changes made during construction, ensuring an accurate representation of the finished work.

Submittals are a critical component of the construction process. They ensure that materials, products, and methods align with the project's specifications and standards, facilitating the approval process and reducing the risk of errors. A well-managed submittal process is essential for maintaining QC, preventing delays, and ensuring the project meets safety, performance, and aesthetic standards.

4.2.2.3 Quality Control Plan (QCP)

A QCP is a comprehensive document designed to guide the project team in ensuring that the construction project is completed according to the required standards, specifications, and regulations. The QCP typically includes the following key elements:

1. **Introduction**
   This section provides a brief overview of the project, including the project description, location, scope, and key stakeholders. It also outlines the purpose and objectives of the QCP, ensuring clarity on the goals of maintaining high-quality standards throughout the project.
2. **Roles and Responsibilities**
   The QCP emphasizes that QC is a collaborative effort, rather than the responsibility of a single individual. This section outlines the roles and responsibilities of the QC team, including the owner, owner's representative, design professionals, construction manager, site superintendent, foreman, and any other member that the team deems relevant based on project specific factors. Each of these team members plays a crucial role in ensuring the project meets the highest standards of quality.

   A specific individual of the construction management team should be designated for conducting and documenting mass timber inspections. Per the IBC, as outlined in the subsequent field inspection section, special inspections of mass timber should be performed in accordance with code. That individual should be trained and have experience relative to wood construction. It is recommended that the individual be predominantly onsite and take an active role of conducting site visits during and after erection. Superintendents, field engineers, QC engineers, and project managers may be the best suited to conduct these inspections at the appropriate intervals.
3. **Quality Standards and Specifications**
   This section specifies the applicable codes and regulations that govern the project, including local building codes, the International Building Code (IBC), International Organization for Standardization (ISO) standards, American Society for Testing and Materials (ASTM) standards, and fire safety codes like the National Fire Protection Association (NFPA) Fire Code. These standards ensure compliance with local, national, and international requirements.
4. **QC Measures**
   This section outlines the processes that will be implemented to ensure that the work conforms to the drawings and specifications. It includes the following key elements:

   - **Inspection and Testing:** Processes for inspecting and testing materials and workmanship to ensure they meet specified standards.

- **Subcontractor Management:** Procedures to monitor subcontractors and ensure their work complies with quality expectations.
- **Non-Conformance Management:** Processes for identifying, correcting, and documenting any non-conforming work, ensuring timely resolution.
- **Communication:** Establishment of regular meetings and schedules to review the status of project quality, facilitating continuous improvement and clear communication among the project team.

5. **QA/QC Documentation**
   This section outlines the procedures for handling and storing key QA/QC documentation. Documents such as inspection reports, test reports, non-conformance reports (NCR), workmanship reports, material certificates, data sheets, and audit reports should be properly managed and maintained for future reference and regulatory compliance.
6. **Competency and Training**
   This section defines the qualifications and experience required for all workers involved in the project. It also specifies any necessary training or certifications for workers and supervisors to ensure they understand and adhere to the required quality standards and specifications. This ensures that the entire team is well-equipped to deliver a high-quality finished product.

The QCP is a dynamic and critical document that provides the framework for maintaining quality on a construction project. It promotes a systematic approach to monitoring, controlling, and ensuring the construction process adheres to the specified standards, contributing to the overall success of the project.

4.2.3 Mass Timber QA/QC Considerations

When applying QA and QC measures to mass timber projects, it is essential to account for the unique characteristics of wood as a material. Due to its specific properties, mass timber demands tailored QA/QC practices to maintain structural integrity, durability, and safety. The following are key QA/QC considerations for mass timber construction:

- **Lumber Sourcing**: The timber used in mass timber components must adhere to strict quality standards. To support sustainable sourcing practices, it is essential to use wood that is certified through reputable programs. Certification programs such as the Forest Stewardship Council (FSC), Sustainable Forestry Initiative (SFI), and Programme for the Endorsement of Forest Certification (PEFC) ensure that wood products are responsibly and sustainably sourced.
- **Grading and Certification:** Timber used in mass timber products must adhere to established grading standards set by organizations such as the American Lumber Standard Committee (ALSC), the National Lumber

Grades Authority (NLGA), or other relevant national bodies. The QC process should also include thorough inspections for visible defects, and manufacturers must apply proper grading techniques when selecting the lumber to use in producing the mass timber components. In North America, CLT manufacturing facilities must be certified under ANSI/APA PRG 320 to produce CLT. This standard specifies that CLT can only be made from softwood species recognized and approved by the ALSC. Furthermore, CLT panels produced in North America must use adhesives that comply with the ASTM D7247 standard.

- **Moisture Content**: Wood is a hygroscopic material, meaning it absorbs and releases moisture based on environmental conditions. Effective moisture content control is crucial to prevent problems such as warping, shrinkage, swelling, or mold growth. The target moisture content for most mass timber products is typically around 12 ± 3%, although it may vary depending on climate, wood species, and intended use. Therefore, the lumber used in mass timber manufacturing must be kiln-dried to achieve the recommended moisture content. Moisture levels should be closely monitored throughout both the production and construction phases to ensure they remain within acceptable tolerances.
- **Accuracy in Production:** Mass timber elements must be fabricated to precise dimensions to ensure proper fit during assembly. Any deviations beyond specified tolerances can lead to inefficiencies on-site, including the need for additional work and rework, as well as potential structural and performance issues.
- **Transport, Handling, and Staging:** Mass timber components should be carefully protected during transportation and staging to avoid damage, such as scratches, dents, moisture exposure, chemical spills, or staining. Proper lifting techniques must be used to prevent damage to the edges or surfaces of the timber.
- **Code Compliance:** Mass timber construction may require special consideration for fire resistance and seismic performance. It is essential to ensure that both the design and materials meet applicable building codes and standards, such as the IBC, along with local fire and safety regulations.
- **On-Site Assembly Considerations:** Site conditions must be suitable for the installation of mass timber components. Temperature and humidity should fall within acceptable ranges to prevent damage from extreme moisture fluctuations. During construction, the wood should be protected from excessive exposure to rain and snow to avoid unwanted moisture absorption. Each timber element should be visually inspected for defects before installation, with any damaged or out-of-specification components replaced or repaired. Additionally, mass timber components must be correctly aligned and securely fastened according to the design. Proper connection details, including alignment and correct torque of fasteners, are crucial to ensure accurate load distribution and maintain the structural integrity of the assembly.

The implementation of rigorous QA/QC procedures is essential to ensuring the long-term performance and safety of these structures. Effectively managing the factors outlined in this section will contribute to the creation of high-quality, durable mass timber buildings that meet both performance standards and sustainability objectives.

4.2.4 Field Inspection of Structural Mass Timber

To ensure QC standards are met, construction teams must take an active role in inspecting mass timber components during the erection phase. Once the material passes inspection at the manufacturing facility, it is loaded and shipped by the preferred method—typically by truck or rail. Field inspections start upon delivery from the manufacturer.

4.2.4.1 Inspecting Mass Timber Materials on Delivery

Upon receiving a shipment, construction teams should thoroughly inspect the load for any signs of damage during transit. It is crucial to document any missing components, types of damage, and the number of affected items before accepting the delivery. The manufacturer and driver should be notified immediately, as many manufacturers have specific reporting windows. Failure to report within this timeframe may limit their responsibility for damages. Any panels that are visibly damaged or compromised should not be installed and must be removed from service.

Mass timber products are sensitive to fluctuations in moisture content, which can lead to swelling, shrinkage, surface gaps, or end checking. Checking, a natural characteristic of wood drying, typically appears as longitudinal cracks or separations in the fibers. According to the National Park Service's Technical Bulletin 08-02, checking in glulam timber is a result of drying and occurs when different parts of the wood dry at different rates, leading to fiber separation. To minimize the impact of this natural process, AITC 111 recommends a gradual seasoning period at moderate temperatures, which aids in material acclimation and helps reduce the occurrence of gaps and cracks. While these effects are typical during the seasoning process, they generally do not compromise the structural integrity of mass timber components.

4.2.4.2 Offloading Procedures

Proper unloading of mass timber materials is essential to avoid damage. SmartLam (2020) recommends unloading panels either in a flatwise position or tipping them into a vertical position prior to lifting. Incorrect unloading techniques can subject panels to high stress, especially those with window, door, or mechanical penetrations, which are weaker than solid panels. Improper handling may result in irreparable damage and compromise the structural integrity of mass timber components.

4.2.4.3 Handling Equipment

Construction teams should use equipment that is appropriately sized for the panels. The weight of CLT panels depends on their thickness and grade, and manufacturers can provide specific weights for each piece to ensure the correct equipment is used. Panels should always be lifted from their center of gravity to avoid causing damage. Determining the center of gravity may be challenging, so it is recommended to contact the manufacturer for individual drawings that indicate the proper lifting points.

For equipment with forks, the forks should extend at least ⅔ of the panel's width. To prevent panel damage, teams should use plywood blocking and foam to protect the surfaces. It is not advisable to use unprotected cranes or slings around panels, as they can scrape or damage the edges.

4.2.4.4 Storage Procedures

If mass timber components are not being installed directly from the truck and are instead stored onsite or offsite, proper storage procedures must be followed to prevent damage. Panels laid out on the jobsite can be easily damaged, and quality teams should ensure adequate protections are in place. Once shipments are accepted, the responsibility for maintaining the structural and aesthetic integrity of the materials falls to the construction teams or installers. Proper handling and storage are essential for preserving quality and reducing the risk of damage.

Panels should be stored on blocks that are 6″–12″ high, following the spacing requirements shown in Figure 4.1. The staging area should be level, clean, well-drained, and have adequate airflow to prevent fungal growth. Mass timber components should never be walked on, touched with soiled hands or equipment, or stored near standing water.

In addition to proper staging and stacking, mass timber components should be protected with wraps and coverings to mitigate handling damage and weather exposure. Components should be covered with moisture-resistant material as soon as possible. Alternatively, teams can leave the factory-provided protection in place until the pieces are ready for installation. If additional protection is needed, plastic coverings should be avoided, as they can trap moisture within the panels. Clear coverings should also be avoided, as they allow UV (ultraviolet) rays to potentially damage the material. To ensure proper ventilation, coverings should be slit or punctured at the bottom to allow for drainage and airflow.

These procedures should be included in the site-specific QC plan, and all workers on-site must be made aware of the necessary protections when handling or working near mass timber components. Actively communicating these procedures to contractors helps mitigate potential quality issues associated with the material on-site. Additionally, limiting the duration and quantity of material stored on-site can reduce handling risks. Implementing just-in-time (JIT) strategies not only frees up jobsite space but also minimizes the risk of damage and enhances efficiency during the erection process. Teams should aim to offload every shipment directly into its designated location for installation.

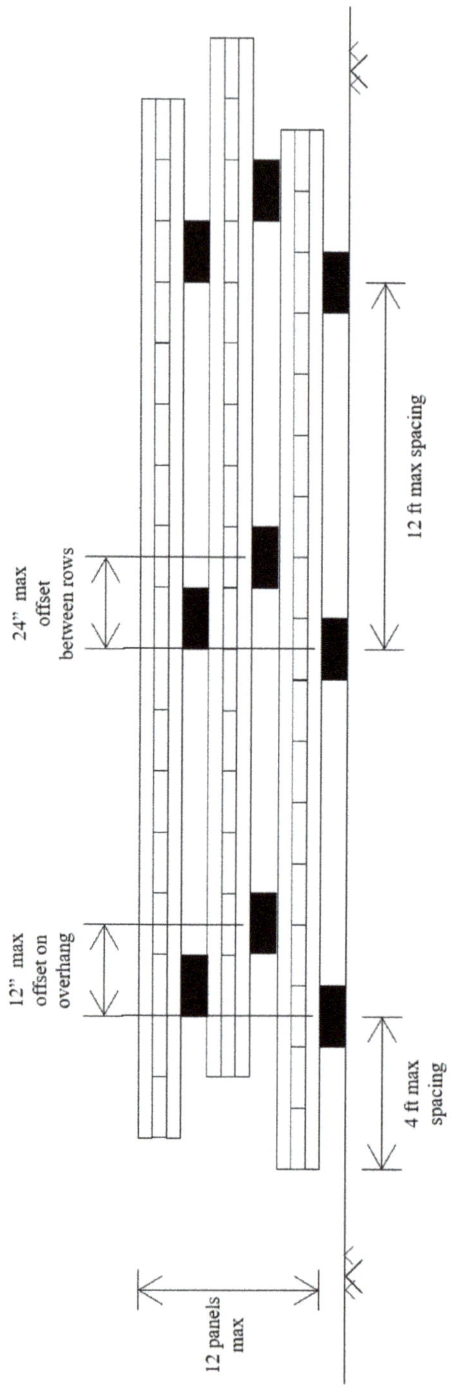

FIGURE 4.1 Proper Blocking Spacing for On-site CLT Staging and Storage.

4.2.4.5 Erection and Installation Inspections

Throughout the acceptance and distribution of mass timber materials, whether stored on-site or installed in their structural location, onsite QC processes must be consistently implemented. These measures should continue post-installation, with teams actively engaged in protecting mass timber components.

It is suggested that teams follow the IBC's recommended inspection practices for mass timber. The 2021 Edition provides recommended inspections for teams to perform. Located in Chapter 17, table 1705.5.3 lists the required special inspections of mass timber construction components. International Code Council(n.d.) defines mass timber construction as structural elements of Type IV construction primarily of solid, built-up, panelized, or engineered wood products that meet minimum cross-section dimensions of Type IV construction (International Code Council (ICC), n.d.).

During the construction process, there are specific items that construction and quality teams should be looking for. This section outlines those specific items and their importance to achieving the quality expectations of mass timber elements. 2021 IBC Table 1705.5.3 outlines each critical area of physical inspections that should take place (see Table 4.1). This table applies to mass timber elements in

TABLE 4.1
Required special inspections of mass timber construction

Table 1705.5.3 Required special inspections of mass timber construction

Type			Continuous special inspection	Periodic special inspection
1	Inspection of anchorage and connections of mass timber construction to timber deep foundation systems		–	X
2	Inspect erection of mass timber construction		–	X
3	Inspection of connections where installation methods are required to meet design loads			
	Threaded fasteners	Verify use of proper installation equipment	–	X
		Verify use of pre-drilled holes where required	–	X
		Inspect screws, including diameter, length, head type, spacing, installation angle, and depth	–	X
	Adhesive anchors installed in horizontal or upwardly inclined orientation to resist sustained tension loads		X	–
	Adhesive anchors not defined in preceding cells		–	X
	Bolted connections		–	X
	Concealed connections		–	X

Source: ICC (2021).

Types IV-A, IV-B, and IV-C construction. Teams should implement this table, procedures for conductions, the inspections, and the documentation process to ensure compliance with the IBC section. In addition, this should be communicated to the erection contractor, ensuring their participation and compliance.

4.3 MASS TIMBER REGULATIONS AND STANDARDS

The International Code Council (ICC) is the largest global association of building safety professionals. The ICC develops trusted codes and standards that are widely adopted worldwide, ensuring safe and high-quality built environments for all stakeholders. Established in 1994, the ICC was formed through the merger of three organizations that had previously developed separate sets of model codes in the U.S.: Building Officials and Code Administrators International, Inc. (BOCA), the International Conference of Building Officials (ICBO), and the Southern Building Code Congress International, Inc. (SBCCI).

The IBC is a comprehensive set of regulations and standards developed by the ICC. It provides guidelines for the construction, design, and safety of buildings and structures to ensure public health, safety, and welfare. The IBC addresses a wide range of building requirements, including structural integrity, fire safety, accessibility, energy efficiency, and mechanical systems.

The IBC is widely adopted by state and local governments across the United States and around the world, serving as the foundation for building regulations. It is updated regularly to reflect advances in building technology, safety practices, and environmental considerations.

Key areas covered by the IBC include:

1. **Building Occupancy and Use**: Classification of buildings based on their function (e.g., residential, commercial, industrial).
2. **Fire Safety**: Fire-resistant materials, fire exits, sprinkler systems, and other measures to protect occupants in the event of a fire.
3. **Structural Design**: Requirements for materials, load-bearing capacity, and safety against hazards like earthquakes, wind, and snow.
4. **Accessibility**: Standards to ensure buildings are accessible to people with disabilities.
5. **Energy Efficiency**: Guidelines for energy conservation in building systems like heating, cooling, and lighting.
6. **Plumbing, Electrical, and Mechanical Systems**: Requirements for safe and efficient operation of essential systems in the building.

4.3.1 Mass Timber in the IBC

The IBC classifies construction types based on the materials used and the fire resistance of the building elements. These classifications are designed to determine the appropriate building standards for safety, fire protection, and structural integrity. The IBC defines five primary construction types:

1. **Type I (Fire-Resistive)**:
    - Buildings are constructed with non-combustible materials, such as concrete and steel.
    - These buildings offer the highest level of fire resistance and are typically used for high-rise buildings, large commercial spaces, and critical infrastructure.
2. **Type II (Non-Combustible)**:
    - Similar to Type I, buildings are primarily made of non-combustible materials like steel and concrete.
    - However, Type II construction may have less fire resistance than Type I and is commonly used for mid-rise buildings, schools, and warehouses.
3. **Type III (Ordinary)**:
    - The exterior walls are typically made of non-combustible materials, but the interior structure (such as floors and roofs) is made of combustible materials like wood.
    - This type often applies to low-rise commercial buildings and residential buildings.
4. **Type IV (Heavy Timber)**:
    - Buildings feature large, solid wood members that are heavily timbered.
    - While the structure is made of wood, the fire resistance is provided by the thickness and mass of the timber, which resists combustion.
    - This type often applies to schools, churches, warehouses, and certain commercial buildings.
5. **Type V (Wood Frame)**:
    - This is the most common construction type, where both the exterior and interior are made of combustible materials, such as wood.
    - Type V construction typically applies to residential buildings, small commercial structures, and single-family homes.

Each construction type is assigned a fire-resistance rating that determines how long building elements must withstand a fire before failure, ensuring the safety of occupants and emergency responders. Teams should familiarize themselves with the construction type relevant to their project and refer to the IBC code for specific requirements applicable to their project.

In 2020, the American Wood Council and the ICC co-authored and published *Mass Timber Buildings and the IBC*, marking the first comprehensive discussion of provisions for mass timber construction within the IBC. Cross-laminated timber (CLT) was introduced in the 2018 IBC revision, which laid the groundwork for significant updates in the 2021 revision, paving the way for broader acceptance and application of mass timber in construction.

4.3.1.1 2021 IBC Inclusions

The 2021 IBC introduced three significant updates that directly relate to mass timber construction. These changes aimed to facilitate the use of mass timber as a viable and safe building material for taller, more complex structures, while addressing fire safety and structural integrity.

1. **New Construction Types for Mass Timber**: The 2021 IBC introduced three new construction types (Type IV-A, Type IV-B, and Type IV-C) designed to accommodate taller and larger mass timber buildings (see Figure 4.2). The fourth type (Type IV-HT) was the existing heavy timber. These new classifications allow for increased height, more stories above grade, and larger allowable areas than were previously permitted for heavy timber buildings.
    - **Type IV-A**: This construction type allows for mass timber buildings in residential and business occupancies up to 18 stories tall. Buildings classified as Type IV-A must have their structural elements fully protected with non-combustible materials.
    - **Type IV-B**: Type IV-B permits mass timber buildings for occupancies up to 12 stories tall. In these buildings, the structural elements are mostly protected with non-combustible materials.
    - **Type IV-C**: This type allows mass timber construction for occupancies up to 9 stories tall. In Type IV-C buildings, most of the structural elements are left unprotected. However, non-combustible protection is

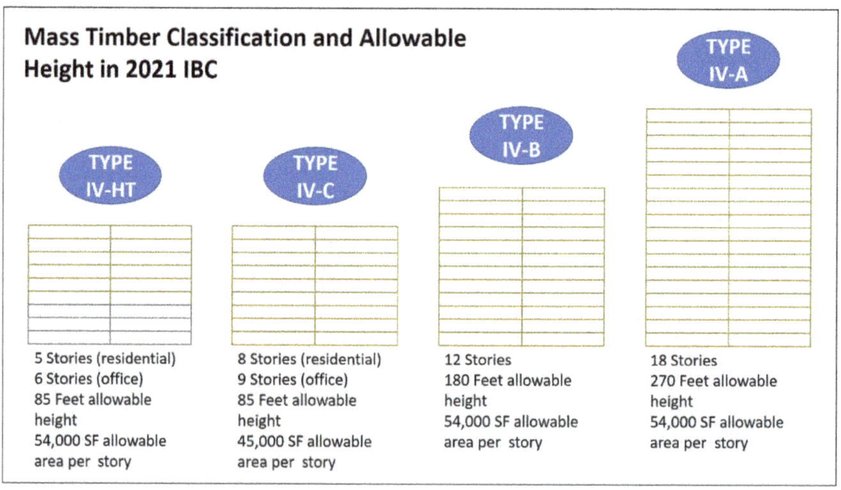

FIGURE 4.2 MT Classification and Allowable Height in the 2021 IBC.

still required for concealed spaces, shaft walls, and the exterior surface of exterior walls.

2. **Special Inspection Requirements**: The 2021 IBC added specific special inspection requirements for the anchorage and connection of mass timber structural elements. These measures ensure the safety, stability, and integrity of mass timber connections and anchorages, which are critical for the performance of tall mass timber buildings.
3. **Enhanced Fire Safety Provisions**: The 2021 IBC includes stricter fire safety measures for mass timber buildings. These include specific requirements for fire-resistance ratings of structural components, the use of fire-resistant coatings, and limitations on the exposure of mass timber elements.

4.3.1.2 2024 IBC Inclusions

The 2024 IBC builds on the progress made in the 2021 version by further expanding the use of mass timber in building construction. A significant change in the 2024 IBC pertains to the mass timber exposure allowances for building type IV-B as outlined below:

- In the 2021 IBC, at least 80% of ceilings and integral beams were required to be covered by non-combustible protection. The 2024 IBC removes this limitation, allowing for the exposure of ceilings and integral beams up to 100% of the fire area or dwelling unit area.
- Like the 2021 IBC, walls and attached (integral) columns are permitted to have exposure up to 40% of the fire area or dwelling unit area.
- Similar to the 2021 IBC, where both exposed walls and ceilings are present, the sum of the ratios (actual exposed area divided by allowable exposed area) must not exceed 1.0.

4.4 WATER MITIGATION

Effective moisture management is critical for every mass timber project, ensuring both concealed and exposed components maintain structural integrity and aesthetic quality. Internal moisture content and external exposure significantly impact performance, and failure to address these factors can lead to costly overruns, schedule delays, and quality issues.

To mitigate these risks, project teams must develop a comprehensive moisture control plan early in the pre-construction phase. Collaboration among all stakeholders including the client, design team, and manufacturer is essential, as moisture management strategies vary in cost, design implications, and scheduling requirements. A well-coordinated approach benefits all parties and enhances project success.

Each stakeholder plays a role in minimizing moisture exposure, with the construction manager responsible for overseeing and enforcing the plan. The following section outlines the effects of moisture, available management strategies, and essential planning considerations to ensure effective moisture control throughout the project.

By implementing a thorough and proactive moisture management plan, project teams can successfully deliver mass timber buildings free from moisture-related issues, regardless of climate. This section provides best practices for moisture mitigation during design and construction, along with additional resources for further guidance.

4.4.1 The Effect of Moisture on Mass Timber

Teams need to understand the potential negative impacts of moisture on mass timber products. The outdoor environment varies based on the project's location, site conditions, and time of year, while the indoor environment is often influenced by building use and occupant behavior. As a result, both the manufacturer and designer must consider moisture mitigation strategies early in the design process.

The most significant form of water exposure for mass timber is liquid water, primarily from rain, but also from other sources including but not limited to snowmelt, runoff, and construction byproducts such as concrete pouring, saw cutting, and plumbing testing. Other moisture loads, such as air, heat, and water vapor, can arise from differing indoor and outdoor conditions, which also contribute to moisture exposure.

Exposure to moisture, especially heavy water, can cause wood to shrink and swell, leading to dimensional changes, gaps, and panel collisions. Rapid moisture exposure may also result in checking (cracking). Additionally, moisture accelerates the degradation of metal connectors and fasteners, promotes decay, and fosters microbial growth. If not properly dried, wood can retain moisture for extended periods, further increasing the risk of damage.

Several factors influence wood's water absorption, including exposure duration, species, and grain orientation. For example, the end grains of CLT panels absorb significantly more moisture than surface grains.

The most common form of water damage in mass timber structures is aesthetic, directly linked to the extent of moisture exposure. Reducing moisture exposure minimizes the risk of fungi, decay, corrosion, and structural damage (Figure 4.3). By controlling moisture, project teams can ensure that any resulting damage remains largely cosmetic.

Overall, the key concerns associated with excessive moisture include the following:

- **Staining and Discoloration:** Too much moisture can lead to aesthetic impacts in the form of dark spots, uneven coloring, and overall discoloration of the mass timber product.
- **Warping and Swelling:** Absorption can lead to warping, uneven surfaces, and can compromise the structural integrity of the product.

FIGURE 4.3 MT Moisture Damage.

- **Fungal Growth:** While a rarer condition, excessive moisture, with prolonged exposure can lead to mold and fungal growth. Increasing the cost of remediation.

While these risks are concerning, they can be effectively mitigated through a comprehensive moisture control plan. With proper planning, project teams can successfully manage these risks, achieving high levels of success with little to no moisture-related impacts.

4.4.2 Environmental Loads and Climate Considerations

Teams should consider geographical location and climate factors during the construction phase, as these can directly affect schedule, cost, and quality, influencing overall construction risks. It is important to evaluate the project's relative humidity and expected rainfall, particularly during construction. National Oceanic Atmospheric Administration (NOAA) and other weather organizations provide data on weather patterns that can help project teams assess the risks associated with constructing buildings in regions prone to high humidity and rainfall during specific periods.

Rainfall totals should be factored into the site-specific moisture control plan (Figure 4.4). If rainfall and humidity levels are high during the erection period, more robust moisture mitigation techniques will be necessary.

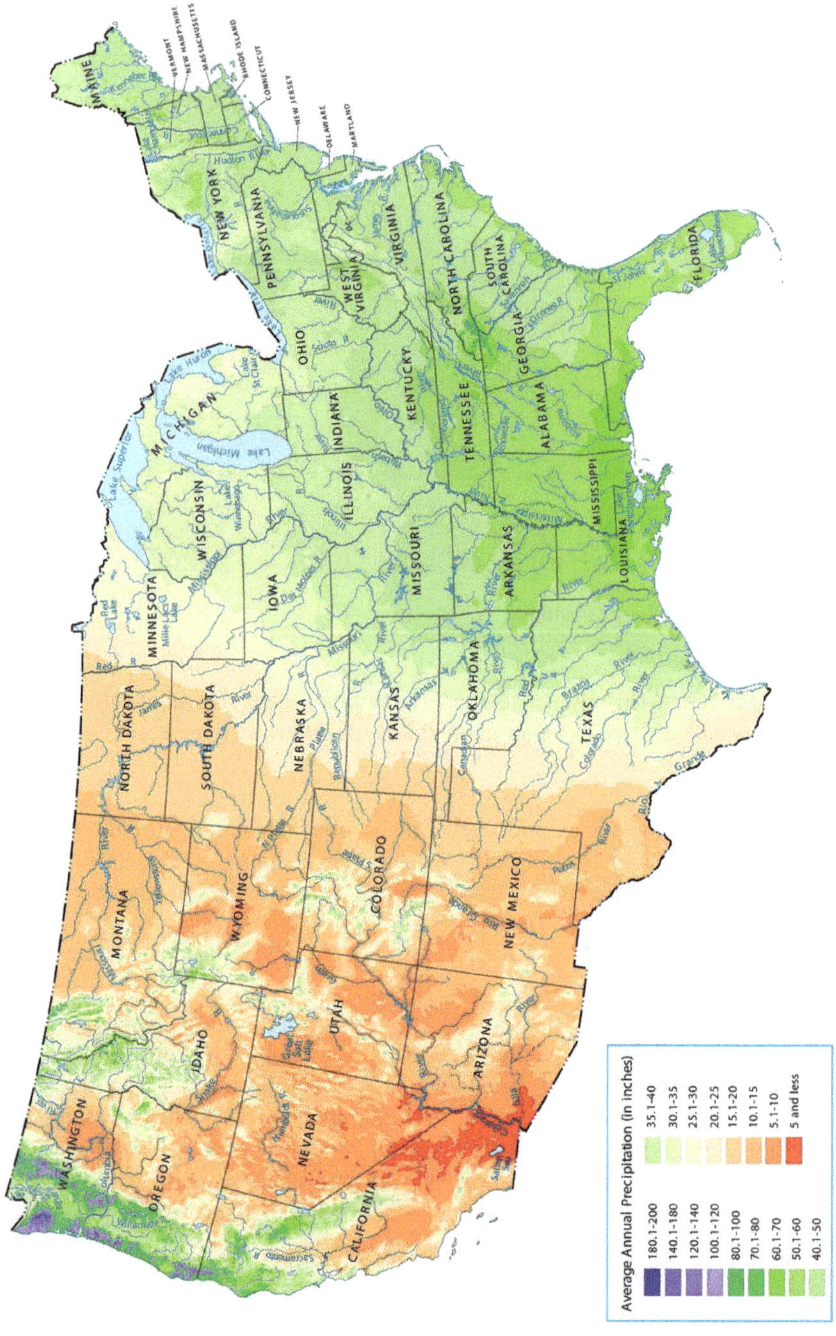

FIGURE 4.4 4 Rainfall Averages for Geographic Regions across the United States (Courtesy of GIS Geography).

4.4.3 Developing a Moisture Control Plan

It is never too early for project teams to develop a comprehensive moisture risk mitigation strategy to protect mass timber structures from water damage. A site-specific approach is essential, requiring construction teams to anticipate potential moisture exposure and intrusion. Likewise, design teams must carefully address water mitigation details, including envelope assembly, flashing, window systems, roof selection, and other protective measures.

This section integrates best practices from builders, architects, developers, and building science professionals. Resources such as *Moisture Risk Management Strategies for Mass Timber Buildings* (RDH Building Science Inc., 2022) offer valuable geographical data and best practices across the United States, providing insights beyond the scope of this text.

It is highly suggested that teams lay out the moisture management process in three steps:

- Step 1: Conduct moisture risk assessment for mass timber.
- Step 2: Develop a moisture quality management plan (MQMP) as part of the project QA/QC plan.
- Step 3: Execute the MQMP.

This section expands on each of these steps, detailing the measures applicable to both the design and construction phases. A recurring theme throughout is the critical role of collaboration among design, manufacturing, construction, and development teams. This integrated approach not only strengthens moisture risk mitigation but also enhances efficiency, reduces costs, and ensures long-term durability.

4.4.3.1 Step 1: Conduct Moisture Risk Assessment for Mass Timber

Companies should expeditiously integrate moisture control plans into their QA/QC playbooks. The first step in this process is conducting a written risk assessment tailored to the site and mass timber structure. This section provides a step-by-step guide for developing and completing this assessment, which should be based on a thorough understanding of site-specific factors affecting the project throughout its design, construction, and occupancy phases. Key factors to consider include elevation, moisture exposure duration, weather patterns, schedule, cost, and the mass timber elements selected. Increased coordination among project principals allows for greater flexibility, enabling the plan to adapt to changing weather conditions and unforeseen challenges.

A critical early step in the moisture risk assessment is determining the project's moisture exposure risk. This is often referred to as the building's moisture exposure level, which indicates the likelihood of structural exposure to wetting events during shipping, storage, installation, or post-construction. Moisture exposure levels are typically categorized according to industry standards and generally correspond to the classifications outlined in Table 4.2.

TABLE 4.2
Moisture exposure levels

Condition	Characteristics	Moisture treatment
Dry Refers to moisture content of 16% or less.	• Indoor, climate-controlled environments with minimal humidity exposure. • Common in residential, office, and commercial buildings.	• No moisture treatment needed.
Moderate Moisture Exposure Refers to conditions with moisture content ranging from 16–20%	• Partially exposed or ventilated areas, such as covered outdoor structures. • Temporary construction exposure to moisture, but with proper drying measures.	• Typically requires additional protection like coatings, membranes, and/or proper detailing.
High Moisture Exposure Refers to conditions with moisture content exceeding 20%	• Prolonged exposure to rain, humidity, or direct water contact. This is found in exposed structures such as mass timber bridges and exposed mass timber facades.	• Typically requires pressure treatment, special coatings, and the use of moisture-resistant adhesives.

The risk of wetting events during shipment, storage, and installation is influenced by factors such as the supplier's location, contract terms, and pre-construction planning. As supply chains become increasingly complex, ensuring product protection during transit is crucial. Protection can take various forms, including wrapping components in vapor-permeable sheets, applying liquid weather barriers, or using weather sealers and coatings. A well-planned design carefully considers material selection, moisture treatment measures, and assembly processes, ensuring an efficient installation that minimizes water exposure and mitigates moisture-related risks. Project teams should consider the risk mitigation options discussed in this section after determining the potential wetting events the project could experience.

4.4.3.1.1 Moisture Monitoring

Monitors are available to assess timber materials during transportation, installation, and use phases. Embedded moisture sensors within the mass timber components continuously collect and transmit data relative to the moisture within the material. This provides actionable data for the project team.

4.4.3.1.2 Applying Liquid Weather Barriers

Liquid-applied weather barriers can be used by project teams to protect mass timber from moisture and water damage. Multiple manufacturers produce liquid barriers that can be adhered to wood, concrete, and other construction materials. Mass timber components must be prepped for application by ensuring the surface is clean, dry, and free of dust and debris. The barrier can be applied with a roller or spray, which cures to form a protective membrane.

4.4.3.1.3 Using Vapor-Permeable Membranes

Mass timber components can also be wrapped in vapor-permeable membranes/sheets to protect against moisture ingress while allowing the wood to breathe. In addition to safeguarding against wetting events, vapor-permeable sheets also offer protection against minor damage, such as dings and dents that can occur during handling. A high-quality wrap can deflect contact with obstacles like metal studs, ladder strikes, dropped tools, and scissor lift bumps.

4.4.3.1.4 Application of Protective Coatings

Teams can choose to have manufacturers apply coatings to protect mass timber against moisture and UV rays, while also impacting the aesthetic finish to meet project needs. For example, Figure 4.5 shows low-VOC coatings that manufacturers can apply at an additional cost. Though wood is strong, it is still susceptible to weathering and biological degradation throughout its lifecycle. Coatings help minimize these effects while preserving the natural attributes of mass timber products.

When considering wetting event mitigation, it's essential to fully understand all available options. Coatings can be applied by manufacturers, offsite, or on-site by local contractors. If applied after shipping, however, additional moisture protection should be provided during transit. Before delving into the specific types and costs of coatings, teams must carefully evaluate the trade-offs.

Trade-Offs in Coating Applications

The primary trade-off is cost, which can increase with the size of the structure. As coatings can be a significant expense, teams must assess if the investment aligns

FIGURE 4.5 5 Low VOC Coating Options from Manufacturers (Courtesy of Timber Pro Coatings USA).

with the overall water mitigation strategy. With the wide variation in coating costs, it's crucial to evaluate their necessity and choose the most suitable option.

Beyond cost, sustainability should also be considered. Coatings containing VOCs or other chemicals may have environmental impacts, so these factors must be considered early, even during the submittal process.

Coatings can also alter the appearance of exposed mass timber. To avoid design conflicts, discussions about coatings should take place during pre-construction, with mock-ups included in the specifications. This ensures that architects, interior designers, and owners can review and provide feedback on the selected coatings.

Factors to Consider When Selecting Coatings

When deciding if a coating is suitable, teams should consider several factors. The coating must provide long-term water protection while maintaining quality throughout the structure's lifecycle. Some coatings require frequent maintenance, meaning they need to be reapplied periodically. It's wise to choose products with longer maintenance intervals for better value.

The compatibility of the coating with the wood species and substrate is also critical. Different wood species have unique characteristics that may affect their interaction with specific coatings. Additionally, environmental factors like UV exposure, human traffic, humidity, and temperature variations must be considered (Dovetail, 2018).

Coating Application Considerations

When determining how and when to apply coatings, consider the benefits of manufacturer-applied coatings in a controlled environment. While this option might have a higher upfront cost, it ensures a uniform application and reduces potential installation issues, saving time and money compared to on-site application.

If the project permits off-site warehousing or protected storage, contractor-applied coatings may be an alternative. Coatings applied in a controlled environment ensure proper adhesion and compliance with specifications.

A hybrid approach is also feasible, such as applying a primer at the factory for initial wetting protection, with topcoats applied onsite. Regardless of the method, all sides of mass timber members, including end grains, must be coated to ensure maximum effectiveness. For instance, all six sides of CLT panels should be fully protected.

For onsite applications, a pre-construction meeting should be held, and application guidelines should be incorporated into the QA/QC plan. Due to the complexity of proper coating application, it is often best to entrust this responsibility to the manufacturer, ensuring consistent quality and adherence to specifications.

Collaborative Decision-Making

Once the factors influencing coating application are understood, collaboration among the design team, construction management firm, and owner is essential. This coordination is crucial for selecting the right protection strategies and ensuring coatings are applied at the correct stages of the project. Considerations such as protection options, travel distance, transit time, and site storage must be addressed. Higher levels of manufacturer-applied protection come at a premium, so teams must weigh whether the additional cost aligns with project requirements. Alternatively, clear coat spray applications that repel water can be applied directly by the manufacturer or combined with other protective measures for enhanced durability.

There is no one-size-fits-all solution for mass timber moisture protection. However, teams can mitigate risks by carefully evaluating the variables discussed in this chapter. The chart provided by the American Institute of Architects (AIA) below outlines the pros and cons of various coating options (Table 4.3).

TABLE 4.3
Moisture protection treatments for mass timber

Treatment	Benefits	Limitations	Notes
Fully adhered vapor permeable sheet	• Generally keeps bulk water out • Water will not travel far in case of breaches • Generally spans checking or cracks	• May be slick for walking (depending on material used) • Liquid water sitting on upper surface for extended periods of time will migrate through as vapor	Some degree of vapor permeability is preferred, but this should not be considered waterproofing
Liquid-applied weather barrier	• Generally keeps bulk water out • Water will not travel far in case of breaches	• Water and temperature application limitations • Application in factory is preferred • Limited ability to span cracks or checks	Options have ranges in vapor permeability, including: • Silicone • Urethane • Silyl-terminated polyether (STPE) • Acrylic
Wood sealers	• Inexpensive • Can be sprayed • Maintains wood appearance	• Weather and temperature application limitations • Requires drying time and multiple coats	• Benefits are similar to liquid treatment, but not as robust
Edge treatment only	• Less labor time to install	• Only shields the end grain • Requires more diligence to keep surfaces dry	• Tapes or sealants

Source: AIA (2019).

4.4.3.2 Step 2: Develop a MQMP as Part of the Project QA/QC Plan

Having assessed the project-specific risks associated with moisture, the team should develop a MQMP to be implemented throughout the life of the project starting from design, through manufacturing, construction, and use. The plan needs to be reflected in the written QA/QC plan that is made part of the contractual obligations of all parties involved. The MQMP should include:

1. **Pre-Construction Phase**
   - Design Considerations:
     - Overhangs, flashing, ventilation, and drainage details to minimize water exposure.
     - Selection of appropriate coatings or sealants for protection.
   - Material Handling & Storage:
     - Shipping protection (e.g., shrink wrap, tarps, or pre-sealed panels).
     - On-site storage guidelines (elevated platforms, waterproof covers, ventilation).
   - Moisture Thresholds & Acceptable Limits:
     - Define acceptable wood moisture content (%MC) levels (typically 12–16% for installation).
     - Set protocols for drying if %MC exceeds limits.
2. **Construction Phase**
   - Site Protection & Sequencing:
     - Temporary covers, tenting, or enclosures to shield wood from rain/snow.
     - Scheduling installation to minimize exposure time.
   - On-Site Moisture Monitoring:
     - Placement of moisture sensors at critical locations (e.g., beam ends, panel cores).
     - Regular manual moisture meter readings.
   - Weather Response Plan:
     - Drying protocols (e.g., fans, dehumidifiers, heaters) for wet components.
     - Procedures for halting work during extreme weather events.
   - Documentation & Accountability:
     - Moisture log tracking readings, weather conditions, and mitigation actions.
     - Assign roles for moisture management oversight.
3. **Post-Construction Phase**
   - Final Moisture Inspection & Testing:
     - Verify %MC before closing walls and ceilings.
     - Address any remaining moisture concerns before occupancy.
   - Long-Term Monitoring Plan:
     - Retain in-place sensors for ongoing tracking.
     - Set maintenance protocols to check moisture levels periodically.

- Response Plan for Future Moisture Issues:
    - Define how to handle leaks, condensation, or unexpected moisture intrusion post-occupancy.

4.4.3.3 Step 3: Execute the MQMP

Moisture mitigation during construction falls into two categories: manufacturer-installed protection and site-installed protection. This section focuses on site-installed protection, though both play a role in the overall protection plan. Site-installed protection can be applied before or after placement and may include treatment of other building materials that interface with mass timber components. As discussed, edge grains are particularly vulnerable to moisture absorption and should be properly sealed.

Once the product is shipped and arrives on-site, the construction manager assumes responsibility for managing the materials. Proper site logistics and storage planning are critical before the product's arrival. While off-site storage, cost, and scheduling are discussed in other sections, it's important to highlight the value of a weather-controlled storage space which helps protect materials from moisture exposure and other potential issues.

Beyond logistics and storage, construction teams play a key role in managing rain events on exposed areas of the building. Several site conditions influence the severity of wetting events, including temporary water diversion, the slope of mass timber panels, and access to temporary roof or floor drains. However, these measures can backfire if not properly managed. For example, drains may clog with construction debris, or pre-graded slopes could be disturbed, leading to water pooling in unintended areas.

Active management of jobsite water conditions should be clearly outlined in the MQMP. Construction managers should incorporate temporary water mitigation requirements into contracts, ensuring subcontractors understand the importance of keeping the structure dry. Setting clear expectations and holding subcontractors accountable are crucial for maintaining QC.

Temporary protection measures are common in construction and can also be applied to mass timber, just as they are used for exterior masonry, painting, and stucco. Temporary roof structures, while more extreme, have successfully protected buildings during construction, highlighting the importance of moisture protection by any means necessary (see Figure 4.6).

Moisture risks extend beyond weather events. Typical construction processes, such as concrete pouring and finishing, introduce moisture that can impact exposed columns and CLT panels. Figure 4.7 illustrates protective plastic wrapping around column bases to prevent water and cement from affecting the finish. These protective measures must be implemented by responsible contractors, with construction managers overseeing water mitigation efforts.

Regardless of the chosen approach, success hinges on clear communication and alignment among stakeholders throughout all project phases. A well-defined

FIGURE 4.6 Temporary Roof Structure (Gasparri et al., 2015).

FIGURE 4.7 Protective Plastic Wrapping.

and agreed-upon moisture management plan increases the likelihood of effective execution and long-term structural integrity.

4.4.3.3.1 Other Factors to Consider During Construction

Exposure Duration

The duration of wetting events significantly influences the extent of potential damage. Exposure time depends on factors such as overhead protection, the speed of constructing subsequent levels, construction delays, and other protective measures in place. To minimize water exposure, teams must closely monitor project schedules, ensuring timely erection and enclosure of the building. Proactive planning and swift execution are essential to mitigating moisture-related risks.

Work on Subsequent Floors

Once the floor is installed, subsequent trades will begin work promptly to keep the project moving forward. The encapsulation of mass timber elements with water-sensitive materials, such as gypsum board, increases the risk associated with wetting events. Once these materials are installed, the need for water mitigation becomes even more critical.

While this challenge exists in other building types, mass timber projects face additional risks due to the application of topping slabs on each floor. With finishes being installed soon after, moisture from concrete pours can seep down, potentially damaging both the timber and other water-sensitive materials.

For best results, refer to the scheduling section to determine the optimal sequence of operations with water sensitivity in mind. Effective pre-planning allows mass timber teams to improve efficiency while minimizing moisture-related risks.

4.4.3.3.2 Post-Construction Exposure Considerations

Wetting events will continue to happen during the occupancy phase of the building's life cycle, and water mitigation measures are critical to ensure longevity. Occupancy phase moisture events like appliance leaks, plumbing leaks, snow and ice melting, and food preparation can all affect the product. Developers should consider this phase when deciding on coatings and maintenance plans, as higher upfront costs could result in long-term value for the ownership group.

4.5 MASS TIMBER FASTENERS AND CONNECTIONS

Connections are a crucial component in any construction project. In mass timber construction, you can expect a range of connections, including timber-to-timber, timber-to-concrete or masonry, and timber-to-steel. Additionally, you may need to connect mass timber elements to light-frame components, which requires a different approach. Within these categories, the specific connections vary depending on the elements being joined, such as floor panels to floor panels, floor panels to wall panels, wall panels to roof panels, beams to wall panels, and panels to steel beams. These variations necessitate a range of connection options, which the design team must carefully select. The choice of connection types is influenced by the following factors:

- **Live and Dead Load Transfer:** Connections should be designed to effectively transfer loads across timber elements, whether experiencing tension, compression, shear, or bending.
- **Movement Due to Expansion and Contraction:** As wood is hygroscopic and responds to changes in moisture content by expanding or contracting, connection designs need to account for this movement.
- **Ease of Assembly:** Prefabricated and easily assembled connections can speed up construction and reduce on-site errors, making them a preferred choice for mass timber projects.
- **Aesthetics:** In mass timber projects, where timber is often left exposed, the design of the connections is crucial to the structure's visual appeal. In some cases, concealed connections are preferred for a clean, seamless finish, while in other instances, exposed connections are chosen to highlight robust steel elements for a more industrial or architectural look.
- **Fire Resistance:** While timber is combustible, it inherently offers fire resistance, primarily due to the "charring effect." However, connections (particularly hybrid connections) pose additional design challenges that must be carefully addressed. When working with mass timber, it's crucial to meet code requirements for fire-resistance ratings (FRR) for both the structural elements and the connections used.
- **Seismic and High Wind Exposure:** In seismic regions, mass timber connections must be designed to withstand the intense forces generated by earthquakes. This necessitates the use of robust and flexible connections capable of handling those loads. Similarly, areas prone to high winds, particularly those with tall, exposed buildings, require connections that can resist wind-induced forces and ensure stability under extreme conditions.

Although there is a wide variety of connection types, proper installation must be accompanied by thorough inspection to ensure that the connections are installed correctly and that they meet the required quality and specifications.

4.5.1 Mass Timber Fasteners

Fasteners are crucial hardware components used to join various elements, including mass timber components. A wide range of fasteners is available for different applications, such as screws, nails, nuts and bolts, adhesives, staples, and rivets. These fasteners come in diverse shapes, sizes, and finishes, including steel, zinc, brass, and aluminum. Mass timber structures are heavily reliant on fasteners to achieve structural integrity, and suppliers are typically consulted early in the design process to offer guidance. For detailed specifications and options, it's important to consult manufacturer catalogs tailored to mass timber applications.

4.5.1.1 Screws

In mass timber construction, the most commonly used fasteners are screws, nails, and GiRods. Screws can be either fully or partially threaded. Fully threaded screws offer high withdrawal capacities and are ideal for reinforcement applications. Partially threaded screws, on the other hand, are perfect for pulling timber members tightly together, often used in situations where head pull-through or lateral loading is a critical design factor. These screws are available in various sizes, head types (such as countersunk, hexagonal, and reverse hexagonal), and tip configurations to suit specific structural needs.

4.5.1.2 Nails

In the fabrication of nail-laminated timber (NLT or nail-lam), nails play a crucial role in fastening the constituent layers of timber together. The type of nail used can vary depending on the specific requirements and application. Smooth shank nails feature a smooth, uniform shaft without any ridges or grooves. The shank remains consistent from head to point, relying on the friction between the nail and the wood to stay in place. Due to their smooth design, these nails provide less resistance to withdrawal compared to other types, making them suitable for applications where holding power is not a primary concern. In contrast, ring shank nails are designed with a shank that is wrapped in small rings or ridges, which are spiraled or grooved around the shaft. The rings significantly increase the friction between the nail and the timber, providing superior holding power and greater resistance to being pulled out. This makes ring shank nails ideal for applications where high tensile strength and long-term durability are required, such as in heavy-duty framing or structural connections.

4.5.1.3 Glued-in Rods (GiRods)

GiRods are specialized reinforcement elements used to enhance the connections between timber components. Typically made from high-strength steel, these rods are embedded into the timber using a bonding agent, such as epoxy or a high-performance adhesive. This creates a strong, durable bond between the timber members that can withstand various loads and environmental conditions. GiRods are particularly effective in reinforcing timber connections where high strength and resistance to tension and shear forces are needed, such as in heavy structural applications or areas exposed to significant stress.

4.5.2 Mass Timber Connections

In their publication, *Mass Timber Connection Index: Optimal Connection Considerations*, Kinder and Kingsley (2024), categorize mass timber connections into three distinct classes based on factors such as cost, complexity, constructability, and fire rating:

- **Class 1: Timber-Only Connections**
  This class features connections that depend solely on mass timber elements and structural fasteners, such as bolts, screws, or dowels. The connection's strength and performance are derived entirely from the timber and fasteners, without requiring additional materials or reinforcements. These connections are often the most cost-effective and easiest to install. However, these connections have loading limitations when compared to class 2 and 3 connections.
- **Class 2: Hybrid Connections with Custom Steel Components**
  Connections in this category combine mass timber with custom-fabricated steel elements, such as plates, angles, or brackets, along with structural fasteners. These hybrid connections offer improved load-bearing capacity and precision, leveraging the strengths of both timber and steel to support more complex or demanding designs. While typically more expensive than Class 1 connections, they provide enhanced performance for intricate applications.
- **Class 3: Connections Using Prefabricated Specialized Connectors**
  This class utilizes pre-manufactured proprietary connectors, typically supplied by specialized vendors. These connectors are engineered for specific applications, ensuring fast and reliable assembly while delivering optimal performance. They enable efficient, precise timber connections, whether between timber elements or in combination with other materials like steel or concrete. These connections are typically the most expensive option and are preferred in projects requiring high strength, fire resistance, and/or hidden connections. Often supported by testing for strength and fire ratings, they provide a reliable solution for demanding specifications.

The WoodWorks Index of Mass Timber Connections is a comprehensive resource offering guidelines and solutions for designing and implementing various timber connections in mass timber structures. It is designed to help engineers, architects, and builders select the most suitable connection systems for different mass timber products, including CLT, glue-laminated timber (Glulam), and NLT (Woodworks, 2024). For detailed descriptions and properties of different connection types—including panel to panel, panel to beam, floor panel to wall panel, wall panel to roof panel, and mass timber to steel, concrete, or light-frame—please refer to the index itself.

4.6 CHAPTER SUMMARY

Ensuring that a client's project meets contractual requirements and quality standards is a fundamental aspect of construction management. Like other critical pillars, a project cannot be considered successful without meeting these quality expectations. In the case of mass timber construction, quality expectations often differ from those associated with traditional structural methods. The visibility of mass timber elements requires multiple QC processes throughout project life cycle, including design, manufacturing, delivery, installation, and post-installation.

This process can be effectively managed by teams that adopt the strategies outlined in this chapter. However, if not properly managed, mass timber components may fail to meet quality standards, presenting significant risks to stakeholders.

It is essential for teams to establish finish expectations early in the design process. Discussions about various finish options should occur before developing a site-specific QC plan and determining associated costs. Construction teams should also focus on proactively addressing quality discussions to minimize potential changes during construction. For example, modifications to drywall coverage can affect the amount and location of exposed timber, which may require negotiations with local building authorities. Delaying these discussions increases the potential for risk. Projects that define and adhere to clear quality standards will enjoy the numerous benefits of high-quality, exposed mass timber construction.

4.7 CHAPTER QUESTIONS

1. What cumulative factors should be considered when determining the potential moisture exposure for a structure?
    a. Exposure duration
    b. Geography, climate, and seasonality
    c. Shipping and storage
    d. Work on subsequent floors
    e. Post-construction, occupancy exposure
    f. All of the above
    g. None of the above
2. Every single mass timber project must implement which of the following, regardless of building type, location, and stakeholders?
    a. a water mitigation plan
    b. site-specific quality control measures
    c. define mass timber finish expectations
    d. all of the above
3. Where are special field inspection requirements for mass timber projects found?
    a. International Building Code (IBC), 2021 edition, chapter 17-table 1705.5.3
    b. International Building Code (IBC), 2023 edition, chapter 17-table 1705.5.3
    c. International Code Council (ICC), 2021 edition, chapter 16-table 1705.5.5
    d. International Code Council (ICC), 2023 edition, chapter 16-table 1705.5.3
    e. None of the above

4. It is suggested that construction teams layout the moisture management process in three steps; please select all steps that apply:
   a. complete a moisture risk assessment for all mass timber components
   b. develop a moisture quality management plan specific to the project and integrate this plan into the overall quality control/quality assurance plan
   c. rely on the manufacturer to protect mass timber components throughout the delivery and installation phase
   d. execute the design and moisture management plan, ensuring stakeholders are held accountable
   e. All of the above
5. The duration of wetting events are dependent on several factors. These include which of the following?
   a. speed of construction
   b. availability of overhead protection
   c. construction delays
   d. project protection measure
   e. **all of the above**
   f. A and C only
6. There are three main factors project teams must consider relative to moisture exposure of mass timber components. These include staining and discoloration, warping and swelling, and fungal growth.
   a. **True**
   b. False
7. Quality assurance processes focus on detecting and resolving issues, while quality control emphasizes preventing quality issues.
   a. True
   b. **False**
8. In this panel-to-panel connection type, the method involves butting adjacent floor panels together, each with a routed surface. A plywood spline is then inserted between the panels, with partially threaded screws or nails used to fasten the spline to both panels, ensuring a secure connection.
   a. steel surface spline
   b. tube connection
   c. **single surface spline**
   d. half-lap joint
   e. screw across a butt joint
9. This joint method involves adjacent floor panels cut with compatible notches that overlap when placed together. The panels are then connected using partially threaded screws or nails, creating a secure and stable connection.
   a. **half-lap joint**
   b. steep surface spline
   c. single surface spline
   d. screws across a butt joint
   e. tube connectors

10. In what year were significant changes reflected in the International Building Code regarding mass timber structures?
    a. 2022
    b. 2023
    c. 2013
    d. 2021
    e. 2019
11. When project teams are considering coatings for exposed mass timber elements, there are several factors to consider. Choose all that apply:
    a. if the coating offers water protection
    b. maintenance frequency
    c. the wood species in which it will be applied
    d. temperature range
    e. **all of the above**
    f. A, B, and C
12. Occupancy phase moisture considerations in mass timber structures include appliance leaks, plumbing leaks, snow and ice melt, and food preparation.
    a. **True**
    b. False
13. There are several factors that influence wood's absorption of water. These include which of the following?
    a. wood species
    b. grain orientation
    c. location
    d. duration of exposure
    e. all of the above
    f. **A, B, and D**
14. The 2021 and 2024 International Building Code stipulates that walls and attached (integral) columns are permitted to have exposure up to _____% of the fire area or dwelling unit area.
    a. 15
    b. 25
    c. 45
    d. **40**
    e. 50
15. Key elements of ensuring site-specific work conform to the drawings and specifications include which of the following?
    a. inspections and testing
    b. subcontractor management
    c. non-conformance management
    d. communication
    e. **All of the above**
    f. None of the above

REFERENCES

AIA. (2019). Mass timber moisture protection considerations for CLT construction. https://aiaseattle.org/wp-content/uploads/CLT-moisture-document-08-13-19.pdf

Dovetail. (2018, October 26). The role of coatings in mass timber performance. Wood Design & Building. www.wooddesignandbuilding.com/role-coatings-mass-timber-performance/

Gasparri, E., Lucchini, A., Mantegazza, G., & Mazzucchelli, E. S. (2015). Construction Management for Tall CLT Buildings: From Partial to Total Prefabrication of Façade Elements. *Wood Material Science & Engineering*, 10(3), 256–275. https://doi.org/10.1080/17480272.2015.1075589

ICC. (2021). Table 1705.5.3 required special inspections of mass timber construction. In 2021 International Building Code (IBC). International Code Council.

International Code Council (ICC). (n.d.). Digital codes. https://codes.iccsafe.org/s/IBC2021P1/chapter-6-types-of-construction/IBC2021P1-Ch06-Sec602.4

International Code Council. (2024, October 31). Tall mass timber provisions represent historic new building code requirements – ICC. ICC. www.iccsafe.org/building-safety-journal/bsj-technical/tall-mass-timber-provisions-represent-historic-new-building-code-requirements/

Kinder, E., & Kingsley, G. (2024). Mass timber connections index: Optimal connection considerations. In *WoodWorks*. WoodWorks. Retrieved February 9, 2025, from www.woodworks.org/resources/mass-timber-connections-index-optimal-connection-considerations/

Nordic Structures. (2020). QC/QA for mass timber construction. In *NS-NT203-US*. www.nordic.ca/data/files/datasheet/file/NS-NT203-US_QC-QA_for_Mass_Timber_Construction_2020-03-27.pdf

RDH Building Science Inc. (2022). Moisture risk management strategies for mass timber buildings. In *rdh.com*. Retrieved February 7, 2025, from www.rdh.com/resource/moisture-risk-management-strategies-for-mass-timber-buildings-v2/

SmartLam North America. (2020). (tech.). Cross-laminated timber delivery, handling and storage. Retrieved from SmartLam North America: www.smartlam.com/wp-content/uploads/2020/05/SmartLam-May-2020-Cross-Laminated-Timber-Delivery_Handling_Storage.pdf.

WoodWorks. (2024). WoodWorks Index of Mass Timber Connections. In *WoodWorks*. Retrieved February 9, 2025, from www.woodworks.org/resources/index-of-mass-timber-connections/

5 Sustainability

"Preservation of our environment is not a liberal or conservative challenge, it's common sense". **Ronald Reagan : 40th President of the United States**

LEARNING OBJECTIVES

After studying this chapter, the reader should be able to:

- Understand the history and theoretical foundation of sustainability.
- Comprehend sustainable forestry practices.
- Recognize mass timber's properties relative to sustainability.
- Understand the impact of circular building economies and its relationship to mass timber.
- Understand third-party assessments used in evaluating mass timber structures.

5.1 INTRODUCTION AND HISTORY OF SUSTAINABILITY

Sustainability is a multi-criteria concept that evaluates systems based on various metrics, including the quality of human life, and is validated through economic, environmental, and social dimensions (Afgan & Carvalho, 2010). The predominant sustainability paradigm (also referred to as the "triple bottom line") represents three interconnected pillars; economic, social, and environmental, which are often depicted graphically as overlapping goals that share resources and create synergy (Purvis et al., 2019), as shown in Figure 5.1. Others depict the stakeholders as people, planet and profit (three P's/PPP) with various visual configurations. Although these representations may vary slightly, they all convey the idea of equal and overlapping stakeholder interests. It is crucial for project teams to utilize this model as a benchmark for assessing their projects.

Historically, these components were viewed as separate entities. The interconnected portrayal of these three stakeholders, however, gained widespread recognition in the 1980s. Historical roots of sustainability are embedded in the

DOI: 10.1201/9781003482543-5

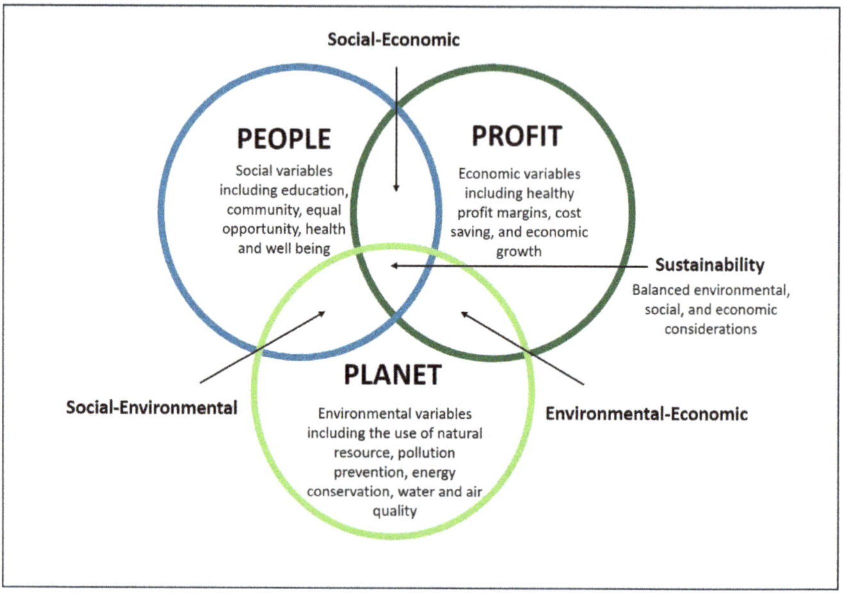

FIGURE 5.1 The Sustainability Paradigm.

concept of sustainable yields which was first introduced in the 17th and 18th centuries by forestry experts in response to the depletion of forest resources across Europe (Purvis et al., 2019). By the late 20th century, dialogues between biocentric preservationists and anthropocentric conservationists were occurring, focusing on sustainable consumption methods that would preserve both nature's intrinsic value and natural resources (Callicott & Mumford, 1997). Following the Industrial Revolution, political economists began exploring the relationship and trade-offs between wealth generation and social justice, laying the groundwork for connecting these concepts (Caradonna, 2014).

The 1970s marked a pivotal decade in the development of the sustainability paradigm. A key event was the establishment of Earth Day on April 22, 1970, following an executive order by President Nixon, just months before the creation of the United States Environmental Protection Agency (Richard Nixon Museum and Library, 2021). In 1975, the Ecology Party's publication of the *Manifesto for a Sustainable Society* further propelled sustainability into mainstream discussions, alongside considerations of the critical factors for sustainable societies. This momentum contributed to the introduction of more efficient standards for state building codes and tax credits for innovative technologies like solar air conditioning and eutectic salt energy storage batteries by the late 1970s (Kibert, 2016).

The combination of federal action, activism, and publications on sustainable practices not only established the foundation for the mainstream use of the term "sustainability" but also expanded its practical applications. However, as energy prices dropped sharply in the early 1980s, the conversation around sustainability stagnated. Nevertheless, later in the decade, several key events reignited the exploration of sustainable principles in practice. Notable milestones included the publication of *Our Common Future* in 1987, the formation of the American Institute of Architects' Committee on the Environment in 1989, and the landmark 1992 United Nations Conference on Sustainable Development. By the early 1990s, the sustainability paradigm had penetrated the construction and development sectors, fueled in part by alarming reports on the built environment's contribution to greenhouse gas emissions (Kibert, 2016).

During the early stages of the sustainability paradigm, a consensus formed that not all pillars were given equal priority, with the economic pillar being prioritized over the social and environmental pillars. In many countries today, the official context for sustainable construction tends to emphasize ecological concerns, often at the expense of social equity issues (CIB, 1999). As the field continues to evolve, the definitions and applications of sustainable construction practices are continually refined, shaping the way project teams must operate.

Recently, monetary incentives have been directly tied to carbon emissions for developers. Additionally, tax incentives linked to sustainability markers have ignited recent efforts to reduce carbon emissions and seek sustainability practices for stakeholders in the built environment. These efforts have a direct impact on cost, schedule, quality, and overall sustainability of projects in which they are applied. As such, they necessitate a shift in management approaches. This chapter will delve into the practical implications of these developments, providing insights into the impact of sustainability certifications on project execution.

5.2 MASS TIMBER AND SUSTAINABILITY

Mass timber offers a multitude of sustainability benefits, which have significantly contributed to its rapid adoption in the construction industry. This section highlights the major sustainability advantages of mass timber, which will be explored in greater detail throughout the chapter.

First and foremost, mass timber is widely recognized for its positive impact on carbon emissions. The growing focus on reducing carbon emissions across various industries, including construction, has led to stricter regulations, the consideration of alternative materials, and a shift toward more sustainable practices. Timber naturally absorbs and stores carbon from the atmosphere during its growth, acting as a carbon sink. When harvested and transformed into mass timber products, much of this carbon remains stored within the wood, preventing its release into the atmosphere. In comparison with carbon-intensive materials like concrete and steel, mass timber offers a significant advantage in reducing overall carbon footprints. Estimates suggest that using mass timber in the construction of commercial structures can reduce emissions by 13–26%, depending on various factors.

Secondly, the use of lumber in mass timber products supports the sustainable forestry industry. Manufacturers source raw timber from suppliers that actively invest in responsible forest management practices. As the demand for mass timber grows, it provides forestry companies with increased resources to invest in the health and sustainability of forests, creating a positive feedback loop between the timber industry and environmental conservation.

Mass timber also offers substantial benefits in terms of construction efficiency. Compared to traditional construction methods, mass timber construction (MTC) can be quieter, faster, and requires less manpower and equipment to erect. These efficiencies contribute to a more streamlined construction process, which in turn enhances the overall sustainability of the built environment. Reduced construction time and resource consumption not only lower operational costs but also lessen the environmental impact associated with lengthy construction phases.

An often overlooked but highly valuable benefit of mass timber is its potential to foster workforce growth. The global market for mass timber is projected to exceed $3.5 billion by 2030, driving the creation of new jobs, manufacturing facilities, and additional construction projects. This economic expansion presents significant opportunities for workforce development, particularly in the sustainable building sector. In addition to its economic advantages, mass timber has a profound effect on the well-being of the people who occupy these buildings. Studies have shown that occupants of mass timber structures experience improved productivity, reduced sick days, and elevated mood levels, thanks to the biophilic properties of wood and its connection to natural environments.

Finally, mass timber is an exceptionally efficient material in terms of waste reduction. Both in production and construction, the processes involved in mass timber manufacturing and building result in minimal waste. The inherent qualities of mass timber lend themselves to a highly efficient construction process, reducing material waste significantly. Additionally, byproducts and leftover materials can be repurposed in a variety of ways. For instance, lower-grade timber components can be used in industrial applications such as matting, while discarded pieces can be milled into sawdust for use in animal bedding products. Non-structural applications, such as temporary roads, scaffolding, and construction platforms, can also be made from surplus timber materials, further minimizing waste and enhancing the overall sustainability of the construction process.

5.3 SUSTAINABLE FORESTRY

One of the key drivers behind the widespread adoption of mass timber in the construction industry is its inherent sustainability. As a carbon-sequestering material, timber outperforms steel and concrete in long-term carbon storage and in reducing emissions at construction sites. Research by architect Robert Cesnik emphasizes that when timber is responsibly sourced, it can be harvested and replenished with minimal environmental impact. Additionally, advancements in engineering technologies and modern building techniques have enhanced the structural predictability and feasibility of mass timber, particularly for mid- and high-rise buildings. To fully appreciate the broader implications of mass timber's rise, it's essential to first examine the origins of the structural supply chain.

Sustainable forestry practices focus on managing forests to meet the current demand for timber without compromising the ability of future generations to access these resources. Without such measures in place, deforestation is likely to accelerate, driven by population growth and international pressures to convert forests into agricultural land for farming and commercial purposes (Ramage et al., 2017). In their *Global warming of 1.5°C* report, the Intergovernmental Panel on Climate Change (IPCC) emphasized the urgency of this issue, noting that sustainable forest management (SFM) strategies—aimed at increasing forest carbon stocks while maintaining a steady timber yield—would yield the greatest emissions mitigation benefits (IPCC, 2018).

Historically, the removal of forest carbon stocks has been viewed as detrimental to the environment. However, recent research suggests that sustainably harvested wood can help mitigate carbon emissions (Lehmann, 2013; Ramage et al., 2017). Forests naturally sequester carbon as trees accumulate biomass, storing this carbon throughout the trees' lifespans. This carbon is only released when the tree dies and decays. By preserving wood in the form of mass timber, carbon storage is extended well beyond the tree's natural life cycle. The responsible sourcing of wood for mass timber production is crucial, requiring the promotion of sustainable forestry practices that prevent deforestation while prioritizing the long-term health and biodiversity of forests. Through sustainable sourcing and forest management strategies, the construction industry can mitigate environmental concerns and ensure that wood resources remain available for future generations.

Despite concerns about global forest depletion, the increased use of mass timber in construction does not need to lead to forest degradation, as long as sustainable harvesting practices are followed (Ramage et al., 2017). In fact, forestry resources have improved in regions such as Asia, North America, and Europe, where 75 billion cubic meters (bcm) of roundwood logs have been harvested since 1990, while forest cover has increased by approximately 1 million hectares per year (Ramage et al., 2017). Europe, which accounts for 90% of global cross-laminated timber (CLT) production—approximately 700,000 m^3 per year—has seen a 90,000 km^2 increase in forest area from 1990 to 2015, driven by sustainable forestry initiatives (Harte, 2017; Ramage et al., 2017). These statistics demonstrate that sustainable forestry practices can coexist with mass timber use and, in fact, can promote forest growth through responsible management.

5.3.1 Deforestation

The growing popularity of mass timber has raised valid concerns about deforestation and the depletion of global forest resources, particularly primary forests (Ramage et al., 2017). Deforestation refers to the clearing or thinning of forests by humans for purposes other than forestry, and it is one of the most pressing issues in global land use. The United Nations Food and Agriculture Organization (FAO) estimates that the annual rate of deforestation is approximately 1.3 million square kilometers per decade, with the highest rates occurring in tropical regions. These tropical forests, which range from lush rainforests to dry open woodlands, are particularly vulnerable. If sustainable practices are not rigorously applied, the rising demand for engineered wood could contribute to forest loss. Sustainable forestry practices, however, represent a long-term, well-planned approach that guarantees the preservation of these resources for future generations (Lehmann, 2013).

The most common form of deforestation in the tropics is slash-and-burn agriculture, or swidden agriculture, where small-scale farmers clear forests by burning them and then grow crops in the nutrient-rich soils fertilized by the ashes. This practice is also used in Southeast Asia, tropical Africa, and the Americas to make way for permanent oil palm plantations. The commercial palm oil industry has rapidly expanded in recent years and led to significant deforestation in Indonesia, Malaysia, and large areas of Africa. Slash-and-burn agricultural methods are often used to clear land for new plantations, resulting in fragmentation of natural forests and loss of habitat for native plants and animals. The expansion of oil palm farming in Indonesia has particularly threatened iconic species such as Bornean and Sumatran orangutans (Pimm, 2025).

One of the most critical areas affected by deforestation is the Amazon Rainforest, the largest remaining block of humid tropical forest. The Amazon spans across Brazil and extends into neighboring countries to the west and north. Studies reveal that approximately 5,000 square kilometers (1,931 square miles) of the Amazon are partially logged each year, with fires consuming an area roughly half the size of the cleared regions annually. This deforestation often results in a fragmented landscape, creating a patchwork of forests and fields or, in more intensive cases, isolated "islands" of forest surrounded by vast expanses of deforested land (Pimm, 2025). However, recent trends in sustainable forestry practices have allowed the region to regain significant hectares of forest.

In recent years, efforts to combat deforestation have included improved forest management practices and the establishment of nature preserves. While these strategies have helped to slow deforestation in some regions, estimates of deforestation still vary depending on the location and the specific methods being employed. The promotion of sustainable forestry practices on a global scale is beginning to show positive results.

While forests may eventually regrow after being cleared and abandoned, this is not always the case, particularly when the remaining forests are highly fragmented. Habitat fragmentation isolates populations of plant and animal species, making it difficult for them to reproduce without facing genetic bottlenecks. In many cases, these forest fragments are too small to support larger or territorial animals. Additionally, deforested areas that are replanted with commercially valuable tree species often lack biodiversity and fail to provide suitable habitats for native plants and animals, including many endangered species (Pimm, 2025).

5.3.1.1 Mass Timber and Deforestation

Mass timber manufacturers source wood from certified mills, ensuring that the timber comes from forests managed according to sustainable practices. SFM is essential for maintaining or improving the health of forests over time. To promote these practices, many countries have enacted national forestry legislation that regulates forest management activities. Certification schemes such as the Forest Stewardship Council (FSC) and the Programme for the Endorsement of Forest Certification (PEFC) offer additional assurances that forests are being managed responsibly (FAO, 2021). These schemes evaluate forest management practices based on a set of environmental, social, and economic criteria to ensure sustainability standards are met.

In Europe, the concept of SFM has become a central pillar in forestry legislation, with several countries taking additional steps to ensure the entire wood value chain—referred to as the "Chain of Custody"—is certified by FSC or PEFC. The Chain of Custody certification ensures that wood and wood products—such as softwood (a key material for MTC) and panel products—can be traced back through the supply chain to verify that they come from sustainably managed forests. This process helps prevent illegal logging, deforestation, and other harmful practices that can damage ecosystems and contribute to climate change. In the case of the Netherlands, making this certification mandatory for (semi)governmental bodies since 2015 has significantly boosted the adoption of sustainable wood sourcing. This requirement has set a strong example for other countries and is aligned with broader European Union goals for reducing environmental impacts. By ensuring that wood products used in construction and other industries are certified, the Netherlands supports both environmental sustainability and responsible consumption. This approach also provides transparency and accountability for consumers and businesses alike, making it easier to ensure that products used in construction and manufacturing don't contribute to environmental degradation or deforestation.

SFM is essential for the long-term health of forests and the critical ecosystem services they provide. Certification schemes like FSC and PEFC help guarantee that forests are managed sustainably, promoting responsible practices worldwide. By using sustainably sourced wood, the MTC industry offers a renewable, environmentally friendly alternative to traditional building materials, contributing to sustainable development.

5.3.2 Reforestation

Reforestation refers to the process of replanting trees on previously forested land that has been degraded though deforestation or natural disasters. According to United Nations Environment Programme, global efforts toward reforestation and tree planting are significant (UNEP, 2019). It is estimated that approximately 1.9 billion trees are planted annually worldwide. This translates to 158 million trees each month, 5 million every day, and approximately 7,000 every minute. This underscores the scale of reforestation efforts occurring globally as individuals, organizations, and governments recognize the urgent need to restore and expand forest cover. The ambitious one trillion trees initiative trees, launched by several organizations, including the World Economic Forum (WEF) and the Trillion Tree Campaign, that aims to plant one trillion trees by 2030 further demonstrates a collective commitment to combat deforestation, enhance ecosystem health, mitigate climate change, and preserve biodiversity. This initiative highlights the global acknowledgment of the essential role trees play in promoting sustainable development and addressing pressing environmental challenges. Such large-scale reforestation efforts aim to restore ecosystems, improve air and water quality, conserve wildlife habitats, and build a more resilient planet for future generations. The rise of mass timber structures has also played a role in shedding light on these vital practices.

As depicted in Figure 5.2, there are five main reforestation techniques, critical for restoring forests and promoting healthy ecosystems.

- **Clear-Cutting Technique:** Involves removing all the trees in a specific area at once. Often used in certain commercial timber harvesting, though it can be controversial due to its environmental impact.
- **Seed-Tree Technique:** A few mature trees are left standing after harvesting to provide seeds for natural regeneration. Common for forests that have tree species that regenerate well from seed.
- **Shelterwood Technique:** Involves removing trees in a series of cuts over several years to gradually create conditions that favor regeneration, while leaving some mature trees behind to protect the younger plants. Often used in areas where gradual regeneration is needed.
- **Group Cuts:** Small, uneven patches of trees are cut to mimic natural disturbances like forest fires or windstorms. This method is used to create a more diverse age structure in the forest, supporting different species and ecosystem functions.

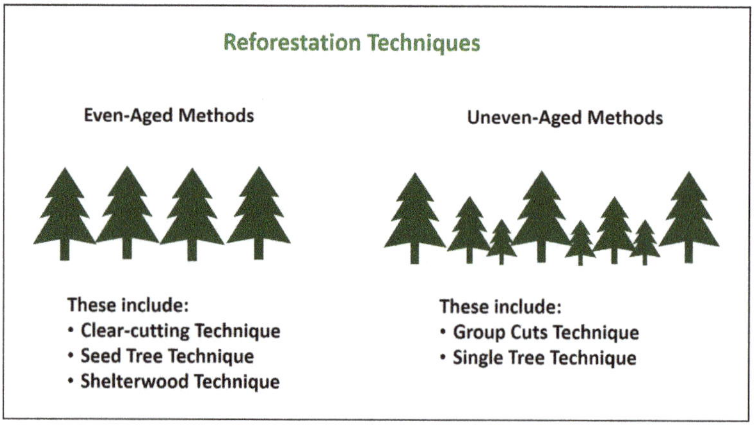

FIGURE 5.2 Reforestation Techniques.

- **Single-Tree Selection:** Involves removing individual trees scattered throughout a forest to create gaps for regeneration while maintaining a relatively continuous forest canopy. Common in forests where a mix of tree ages and species is desired.

Each of these techniques has its place in forest management, and choosing the right one depends on the goals for the forest, the local environment, and the type of trees being managed. While some methods focus more on timber production, others prioritize ecosystem health and biodiversity conservation.

The increasing demand for forest products, when managed sustainably, can drive reforestation efforts and investments in forest management, helping to expand forest carbon sinks and encouraging forest growth and regeneration (Pasternack et al., 2022). However, these benefits can only be realized when sustainability remains the cornerstone of forest management practices. Countries like Austria, Switzerland, Canada, and Finland are leaders in developing and implementing nationwide sustainable forestry policies, ensuring the responsible management of forest resources.

5.3.3 Afforestation

Afforestation is the practice of planting trees on land that has not been covered by forests for a significant period. It is a crucial strategy for increasing global forest cover and addressing deforestation. Through afforestation, forests can help remove carbon dioxide (CO_2) from the atmosphere as trees grow, contributing to carbon sequestration and functioning as a form of carbon removal (American University, 2022). There are five main types of afforestation initiatives identified by Grow Billion Trees (2025)

- **Reforestation of Degraded Lands:** This initiative focuses on restoring deforested or degraded landscapes by planting trees, controlling soil erosion, and facilitating ecosystem recovery. It plays a critical role in enhancing biodiversity, improving soil health, and preserving vital water resources, ultimately restoring ecological balance.
- **Urban Forestry and Greening:** By creating green spaces, parks, and urban forests in cities and towns, this initiative aims to improve air quality, mitigate the urban heat island effect, and foster healthier, more livable environments for urban populations. It enhances residents' well-being and promotes sustainable city development.
- **Agroforestry and Farm Forestry:** Integrating trees into agricultural landscapes, this approach boosts soil fertility, diversifies crop production, and provides farmers with additional income streams. It promotes sustainable farming practices while enhancing ecosystem services, such as water retention and biodiversity.

- **Coastal Afforestation and Mangrove Restoration:** Planting trees and mangroves along coastlines and estuaries protects against erosion, safeguards communities from storm surges, and strengthens coastal resilience to climate change. These vital ecosystems also support biodiversity and serve as natural buffers against extreme weather events.
- **Afforestation in Watershed Management:** Establishing forests in riverbanks, watersheds, and catchment areas helps regulate water flow, reduce soil erosion, and enhance water quality. This initiative is vital for protecting freshwater resources and ensuring the sustainability of water supplies for downstream communities.

The benefits of afforestation go beyond carbon sequestration. Ecologically, the advantages are particularly evident in forest restoration efforts, which focus on creating diverse and resilient ecosystems rather than monoculture tree plantations. Forest restoration projects prioritize the reintroduction of native tree species and work to recreate habitats that support a wide range of plant and animal life. These restored forests offer numerous ecological benefits, including enhanced soil health, water regulation, erosion control, and the provision of critical wildlife habitats.

Additionally, afforestation initiatives can bring social and economic advantages. Forests provide livelihoods, offering sustainable sources of timber and non-timber forest products. They also contribute to local economies through ecotourism and other nature-based activities. Forested areas play an important role in supporting the well-being and preserving the cultural heritage of indigenous communities who have traditionally relied on forests for their subsistence and way of life. Moreover, financial investments in purchasing land and planting trees for production purposes can further boost local economies.

5.4 RAW MATERIAL UTILIZATION

Raw material utilization, also known as material consumption, refers to the amount of raw material used in the fabrication of products or product parts. Effective utilization of raw materials is a key aspect of sustainable design, particularly in relation to the manufacturing process. Maximizing raw material efficiency not only reduces waste but also lowers production costs and supports environmental sustainability. The manufacturing and construction industries depend heavily on raw materials, and modern technologies allow for optimized material use, leading to substantial long-term savings. However, many companies fail to prioritize material utilization, missing opportunities to cut costs without compromising product quality or efficiency.

One of the strengths of mass timber is its efficient use of raw materials sourced from sustainably managed forests. Typically, mass timber is harvested from younger trees, which contributes to a lower environmental footprint. The wood processed by sawmills has exceptionally high yields, with minimal waste (see Figure 5.3).

FIGURE 5.3 Tree Utilization Chart.

Once logs are harvested, they are transported to sawmill laydown yards for processing. To prevent loss from biological deterioration and attacks by insects and fungi, sawmills regularly wet the lumber to maintain a moisture content above the fiber saturation point, typically between 28–30%. This practice helps prevent checking, cracking, and staining of the stored logs. Additionally, sawmills work quickly to process raw lumber in order to reduce the risk of damage to the logs. The next step is debarking, which allows for a clearer view of the log and prepares it for entry into the headrig machine center.

The headrig, often equipped with one or more cutting tools, plays a critical role in mill efficiency. It performs four key functions:

- Determines the grade and geometry of the log
- Orients the log into the correct position for processing
- Transports the log along the processing line
- Completes the required cuts

The yield percentage of sawmills, known as the lumber recovery rate, represents the ratio of lumber output to log input. This percentage can vary depending on the type of wood, log quality, length, age, and processing technology. Since the mid-1920s, lumber recovery rates have seen significant improvements. Initially, only about 40–45% of the log would be converted into usable lumber (Hiziroglu, 2017). However, with the advent of laser and camera scanning technology, modern mills can achieve lumber recovery rates of 70% or higher. Computerized programs, such as the "best opening face" (BOF) algorithm, have also contributed to these improvements.

Sawmill residues are the byproducts or waste left over after the primary lumber processing. These byproducts, including sawdust, wood chips, bark, and shavings, are repurposed for various uses such as energy generation, the production of other wood products, and animal bedding. By effectively utilizing these residual wastes, sawmills can maximize the use of the entire tree, thus improving the yield percentage.

The conversion of sawmill residues into energy is known as biomass cogeneration. Biomass cogeneration is a process that uses organic materials to generate energy. In this process, wood waste is fed into a combustion chamber at an energy plant, where it is burned at high temperatures to create heat. This heat is used to convert water in the boiler system into steam, which is then directed into a turbine to generate renewable energy.

Overall, the modern timber production process is highly efficient. It minimizes waste, maximizes the use of byproducts, and contributes to sustainable practices by repurposing residual materials for energy and other uses.

5.4.1 The Impact of Mass Timber on Climate

The urgent need to combat climate change has prompted policymakers, businesses, and other stakeholders to explore innovative solutions to reduce greenhouse gasses emissions and decarbonize hard-to-abate sectors. One promising approach gaining traction is MTC, which holds significant potential for reducing greenhouse gas emissions. Studies have shown that replacing steel and concrete with mass timber in commercial buildings can decrease emissions by 13–24.5%, considering factors such as manufacturing, transportation, and installation of building materials (Ding et al., 2016).

However, the growing demand for wood products could have unintended consequences, potentially impacting forest carbon stocks, forest health, and other vital social and environmental benefits. It is crucial, therefore, to implement safeguards to ensure that the benefits of MTC outweigh its potential downsides. To address this, the Nature Conservancy (TNC) has launched a global mass timber impact assessment (GMass TimberIA), aimed at evaluating the potential impacts of mass timber on forest ecosystems and emissions associated with the built environment (Schlesinger et al., 2021).

The insights gained from these assessments will help deepen our understanding of the climate implications of mass timber. While mass timber presents a promising climate solution, it is crucial to assess its impact on forests and the broader climate comprehensively. These evaluations can offer valuable information to policymakers, businesses, and other stakeholders, enabling them to maximize the benefits of mass timber while mitigating potential drawbacks.

5.4.2 Sustainable Timber Supply in the United States

Mass timber manufacturers typically source raw materials from sustainably managed forests across the United States. These highly managed forests are known for producing high-quality timber, ideal for engineered wood products. Contrary to common belief, the United States boasts a robust and sustainable lumber economy, largely driven by the efforts of harvesters who prioritize environmental stewardship. These efforts focus on limiting overcutting and degradation while ensuring thorough preplanning to maintain a steady timber supply. As a result, both public and private lands have seen a growing timber inventory, with an estimated 823 million acres of forest and woodland, accounting for roughly one-third of the nation's landscape (U.S. Forest Service, 2022).

Timber harvesting practices vary depending on location, tree type, and age. In southern states, for instance, pine trees are typically considered mature at 25–50 years old (Greene, 2020). This geographic and biological variability ensures that the timber supply is continually evolving, making precise data on timber availability challenging to quantify.

However, the U.S. Forest Service's Forest Inventory and Analysis Program (FIA) tracks these critical data points, providing up-to-date forest health and timber supply statistics. An important aspect of sustainable timber management involves certification programs such as the Forest Stewardship Council (FSC) and the Programme for the Endorsement of Forest Certification (PEFC), which ensure that timber is sourced from responsibly managed forests. These programs help safeguard timber resources, supporting long-term sustainability for future generations.

To further advance the sustainability of the timber supply and promote mass timber adoption, continued investment in research and development is essential. Focused efforts on improving sustainable forestry practices and exploring alternative timber sources can enhance production efficiency, reduce waste, and contribute to biodiversity conservation. These investments are crucial for ensuring that mass timber remains a viable, environmentally responsible building material in the years to come.

5.4.3 Mass Timber and the Circular Building Economy

A Circular Economy offers a systemic approach to addressing global challenges like climate change, biodiversity loss, waste, and pollution. It is built on three key principles, driven by design: eliminating waste and pollution, circulating products and materials at their highest value, and regenerating nature. In the context of mass timber, the circular economy is a framework that emphasizes the sustainable use, reuse, and recycling of timber throughout its lifecycle (Figure 5.4). This approach minimizes waste, maximizes the value derived from timber, and reduces environmental impact, aligning well with the principles of sustainability and resource efficiency.

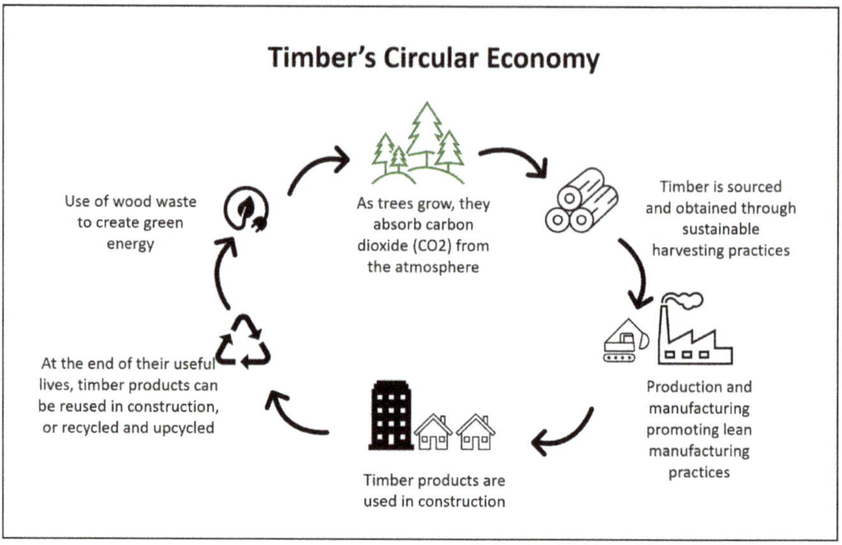

FIGURE 5.4 Timber's Circular Economy.

Sourcing and Sustainable Harvesting: The circular economy begins at the sourcing stage, where timber is harvested from sustainably managed forests. By following responsible forestry practices, such as those endorsed by certification programs like FSC and PEFC, the industry ensures that the trees being cut down are replenished through replanting and natural regeneration, preserving biodiversity and ecosystem health. This sustainable harvesting forms the foundation of the circular timber supply chain, allowing for continuous cycles of timber production.

Production and Manufacturing: Once timber is harvested, it's processed into mass timber products such as CLT, glue-laminated timber (GLT), and other engineered wood products. During this manufacturing stage, the focus is on maximizing material efficiency—minimizing waste, optimizing cuts, and ensuring that as much of the timber as possible is used in the final products. Advanced techniques in mass timber production, like precision milling and digital fabrication, also allow for higher performance and reduced waste in manufacturing.

Construction and Use: In the building and construction phase, mass timber products serve as durable, high-performance building materials with an extended lifecycle. Mass timber is inherently carbon-negative, storing carbon dioxide that was absorbed by the trees during their growth. This long-term carbon storage helps reduce the overall carbon footprint of the building, especially when compared to more carbon-intensive materials like steel and concrete.

Buildings constructed with mass timber can have a long lifespan, reducing the frequency of rebuilding and thus decreasing the demand for new raw materials. Furthermore, mass timber structures can be designed for easy disassembly, enabling components to be reused or repurposed at the end of their initial life. This adaptability makes mass timber an ideal candidate for circular economy principles, as materials can be reclaimed and used again in future projects.

Deconstruction and Material Recovery: At the end of a mass timber building's life cycle, deconstruction becomes a key step in maintaining the circular flow of materials. Unlike traditional demolition, deconstruction focuses on carefully dismantling structures to recover reusable components and materials. High-quality mass timber, such as CLT panels or timber beams, can be reclaimed, refinished, and repurposed in new buildings or other applications, reducing the need for virgin timber and further conserving resources.

Additionally, if the timber can no longer be reused in its original form, it can be processed into other valuable products such as wood chips, biomass energy, or even compost for agricultural uses. This ensures that no material goes to waste and that the timber continues to provide value, even after the building's lifespan has ended.

Recycling and Upcycling: When mass timber products reach the end of their useful life or cannot be reused in construction, recycling and upcycling become critical steps in the circular economy. The wood can be broken down into smaller pieces for use in engineered wood products, or it can be repurposed into biofuels, reducing reliance on fossil fuels. In some cases, upcycling can involve turning used timber into higher-end products, such as furniture, flooring, or decorative materials.

Environmental Benefits: The circular economy of mass timber offers a range of environmental benefits. It reduces the need for new raw materials, lowers carbon emissions, conserves natural resources, and minimizes waste. The use of sustainably sourced timber ensures that forests are preserved and can continue to provide valuable ecosystem services, such as carbon sequestration, water filtration, and wildlife habitat.

Moreover, mass timber structures themselves have a lower carbon footprint compared to conventional building materials, making them an ideal solution for reducing the built environment's impact on climate change.

Incorporating the principles of the circular economy into mass timber production and construction helps create a sustainable, closed-loop system that benefits both the environment and the economy. By focusing on responsible sourcing, efficient manufacturing, long-lasting use, and material recovery, the mass timber industry can significantly reduce its environmental footprint while providing high-performance, sustainable building materials. The continuous flow of materials from one life cycle to the next is the essence of a circular economy and positions mass timber as a leading player in the transition to a more sustainable built environment.

5.5 SUSTAINABILITY ASSESSMENT METHODS

Designers, construction professionals, developers, governments, and consumers are increasingly conscious of the need for products, processes, and systems that are not only economical but also environmentally friendly. Assessing sustainability impacts is achievable, but offers a wide array of challenges. These challenges are often due to the numerous variables, subjectivity, and uniqueness of what is being assessed. No matter the challenges, demand for accurate assessment is clear. This section outlines various systems used to assess sustainability in the built environment along with mass timber's strong relationship with specific methodologies.

5.5.1 Third-Party Sustainability Certifications

In the built environment, buildings are often assessed by third-party sustainability certification systems, especially when stipulated by contract. These systems provide a structured framework for teams to initiate, evaluate, and document a building's sustainability performance. Typically, they rely on a point-based structure, offering teams various options to achieve predefined sustainability levels. Points are weighted to determine the project's overall sustainability rating, reflecting both the building's environmental performance and the associated cost premiums.

Before pursuing third-party certification, the project team must carefully consider their desire, eligibility, and goals for achieving certification. This decision is critical, as it directly impacts the project's cost, schedule, quality, safety, and overall performance. Teams are encouraged to collaborate and communicate thoroughly to weigh the pros and cons of certification, ensuring a well-informed choice.

Once a team has decided its pursuit of certification, inclusion of the requirements of each stakeholder must be reflected in contractual documents. This process should happen as early as possible due to the widespread ramifications in the design and management process. The subsequent section outlines the main certification systems, important attributes, and other factors to consider. There are various systems, but some of the globally recognized ones are briefly described here. While goals and assessment methodologies may vary between rating systems, there are some overarching goals highlighted.

5.5.1.1 Leadership in Energy and Environmental Design (LEED)

LEED is a building certification program administered by the U.S. Green Building Council (USGBC). It focuses on improving building and construction project performance across seven key areas of environmental sustainability and human health. For practitioners, LEED certification is often a critical consideration, particularly during the bidding phase, as third-party sustainability requirements can influence other aspects of construction.

Recent data reveal a strong correlation between mass timber structures and LEED certification. Nearly 39% of commercial mass timber projects are

LEED-certified (Mirando et al., 2023). In contrast, only 1.9% of mass timber projects achieve WELL certification and 3.9% earn net-zero energy (NZE) certification. Other structural systems, such as cast-in-place (CIP) concrete and steel, achieve certification at around 1.5%. This stark disparity demonstrates that mass timber buildings are certified over twenty-five times more frequently than other structural systems.

Given this trend, construction managers should familiarize themselves with the LEED certification process and consider earning LEED credentials for their teams. The LEED rating system not only certifies buildings but also credentials professionals, better preparing them to navigate the complexities of sustainability in the built environment. Professionals in various fields within the built environment can achieve different levels of LEED certification, signaling their commitment to sustainability and highlighting specific expertise in key building types. Teams and individuals interested in learning more about credentialing are encouraged to explore the USGBC website and the range of resources offered by the organization.

Regarding building certification, the LEED system provides flexibility, offering various pathways to certification based on building types and even neighborhoods. Teams are presented with a scorecard containing a point-based system. Certification levels are determined by the number of points a project earns. The four levels, ranked from highest to lowest, are Platinum, Gold, Silver, and Certified. Higher levels of certification reflect a greater investment in sustainability, which is often associated with higher initial costs but lower operating expenses over time. Teams must decide not only whether to pursue LEED certification but also at which level.

LEED is widely regarded as the gold standard due to its market dominance, consumer awareness, and longevity when compared to most other sustainability rating systems. The strong correlation between mass timber structures and LEED certification underscores the need for teams to explore this methodology, particularly if they are unfamiliar with it or wish to gain deeper expertise. Given the data, mass timber projects are highly likely to pursue LEED certification, making it essential for teams to understand and leverage the system effectively.

5.5.1.2 WELL

WELL Certification is a unique and innovative approach to building performance, as it places a strong emphasis on human health and well-being. Created and managed by the International WELL Building Institute (IWBI), this performance-based system ensures that buildings are designed, constructed, and operated in ways that promote the health and comfort of the people who occupy them.

What sets WELL apart from other green building systems like LEED or BREEAM is its distinct focus on human health, rather than solely addressing the environmental impact of the building. WELL specifically targets various aspects of the indoor environment that influence both physical and mental health, including air quality, lighting, acoustics, nutrition, fitness, and the psychological effects of design. By emphasizing these factors, WELL seeks to create spaces that not only minimize harm to the environment but also enhance the overall well-being of occupants. Key components of WELL Certification include:

Performance-Based Metrics: Unlike some systems that are prescriptive, WELL uses performance-based metrics that require measurable evidence of compliance. These metrics help ensure that buildings actually meet the intended health outcomes. For example, a WELL-certified building might need to meet specific air quality standards, ensure access to natural light, or provide spaces that promote physical activity.

Documentation & Testing: To achieve WELL Certification, project teams must thoroughly document their compliance with the required standards and provide evidence of performance through testing and ongoing monitoring. This can include air quality testing, light level measurements, noise assessments, and more. The certification process requires significant effort to validate that the building is not just designed for health but actually performs as expected.

Seven Concepts: WELL focuses on seven key concepts that are integral to the health of building occupants.

- Air: Ensuring clean indoor air quality, including reducing pollutants and improving ventilation.
- Water: Ensuring access to clean, safe, and high-quality water.
- Nourishment: Promoting healthy food options and access to nutritious meals.
- Light: Enhancing natural light exposure and ensuring appropriate lighting for circadian rhythms and productivity.
- Fitness: Encouraging physical activity with features like stair access, fitness centers, or active design principles.
- Mind: Supporting mental well-being through aspects like access to nature, stress-reducing environments, and positive design.
- Sound: Managing acoustics and noise to reduce distractions and create a more comfortable environment.

Levels of Certification: Just like other green building rating systems, WELL offers different levels of certification, ranging from Silver to Gold to Platinum, depending on the number of points the building earns in each category. These points are awarded based on how thoroughly the building meets the performance criteria.

Achieving WELL certification can be a significant undertaking for project teams, as it requires not only thoughtful design and construction but also ongoing maintenance to ensure that performance standards are met over time. However, the benefits for building owners, developers, and occupants are substantial. For building owners, WELL Certification can serve as a marketing tool, highlighting their commitment to the health and well-being of occupants. For tenants, the promise of a healthier environment can lead to greater employee satisfaction, productivity, and retention.

As the understanding of the connection between the built environment and human health continues to evolve, WELL Certification has become an increasingly sought-after credential for both commercial and residential buildings. This growing demand reflects a shift toward prioritizing occupant well-being alongside environmental sustainability, with WELL Certification serving as a key standard for creating spaces that support physical, mental, and emotional health.

5.5.1.3 NZE

NZE buildings are an ambitious and highly impactful goal for sustainable development in the built environment. Essentially, an NZE building produces as much renewable energy on-site as it consumes over the course of a year. This concept isn't new, but technological advancements, growing public demand for sustainability, and stricter governmental regulations have significantly increased its adoption in recent years.

Achieving NZE status requires a multi-faceted approach, integrating several complex systems:

Highly Efficient Building Operations: Achieving energy efficiency begins with implementing advanced technologies and practices that minimize energy demand from the outset. High-performance insulation, energy-efficient windows, LED lighting, smart building systems, and other strategies are essential for reducing the building's energy load. This approach is often referred to as the "reduce" part of energy efficiency—prioritizing the reduction of energy consumption as much as possible before relying on renewable energy generation. By focusing on energy conservation first, buildings can significantly lower their overall energy needs, making it easier to achieve NZE goals.

Renewable Energy Generation: NZE buildings rely on renewable energy sources like solar, wind, or geothermal to meet their energy needs. Solar panels are often the most common form of renewable energy generation used, as they are scalable and relatively easy to install. Geothermal systems, which provide heating and cooling, are another reliable renewable energy source that significantly reduces a building's overall energy consumption.

Holistic Design & Integration: Achieving NZE requires integrating renewable energy systems into the building's design from the very beginning. This goes beyond simply adding renewable energy generation after construction; it involves optimizing the building's form, layout, and materials to work seamlessly with these systems. A holistic approach also incorporates the use of natural systems to enhance efficiency, such as natural ventilation, daylighting, and heat recovery ventilation. By considering these elements during the design phase, buildings can achieve a higher level of energy performance while minimizing reliance on external energy sources.

Environmental Integration: Natural systems are often incorporated into the building's operations to meet energy needs or reduce reliance on mechanical systems. For instance:

- *Harnessing Solar:* Solar panels can meet a large portion of electrical energy needs.
- *Geothermal Wells:* These can provide energy for heating and cooling, reducing the building's need for external energy sources.
- *Rainwater Reuse:* Capturing and reusing rainwater can significantly reduce the water demand for landscaping and operational needs, minimizing the need for potable water.

Achieving NZE status is not an easy or inexpensive process. It requires significant investment, both in terms of capital and time, as well as a strong commitment to sustainability. Designing and constructing an NZE building involves integrating complex systems that require careful coordination among architects, engineers, contractors, and building operators.

However, the benefits are clear. NZE buildings can dramatically reduce operating costs (especially energy costs) over time, offer better occupant comfort, and contribute significantly to reducing carbon footprints. The recent uptick in the number of NZE buildings highlights how advances in technology and shifts in regulatory frameworks are making this goal more achievable and desirable.

Mass timber buildings have a unique advantage in the pursuit of NZE status. Timber, particularly mass timber, is a natural, renewable material that sequesters carbon, which offsets some of the building's emissions during construction. This property, combined with timber's relatively low embodied energy compared to concrete or steel, makes it a strong candidate for helping buildings achieve NZE status. In addition, timber's aesthetic qualities and ability to help regulate indoor air quality and comfort further enhance its appeal in sustainable design.

As the demand for sustainable, high-performance buildings continues to rise, it's likely that we'll see more mass timber buildings play a key role in achieving NZE goals. They offer a unique blend of environmental benefits, energy efficiency, and aesthetic appeal that aligns well with the principles of NZE.

5.5.1.4 BREEAM (Building Research Establishment Environmental Assessment Method)

BREEAM is one of the most well-established and widely recognized green building certification systems in the world. Originating in the U.K. in 1990, it has since been adapted and implemented internationally across a variety of project types. BREEAM's reputation as a leader in sustainability assessment is a testament to its comprehensive, evidence-based approach to measuring a building's environmental impact.

The BREEAM rating system operates on a scale that ranges from Pass to Outstanding, with each project earning points based on its performance in various environmental categories. These categories include energy use, water consumption, waste management, and materials sustainability. BREEAM places particular emphasis on achieving improvements in environmental impact across the life cycle of a building, from design to construction and operation.

What makes BREEAM stand out is its focus on environmental impact, with different categories being weighted based on their importance and potential to reduce a building's overall carbon footprint. For example, energy efficiency and water management often have a higher weight because of their significant role in reducing operational environmental impact. This ensures that the rating system directs focus on the most critical aspects of sustainable design and construction.

Regarding mass timber, it's interesting to note that BREEAM has been particularly relevant to this type of construction in the UK. Mass timber projects are increasingly aligned with BREEAM due to the system's attention to the environmental benefits of using renewable, low-carbon materials like timber. Since mass timber often has a more favorable environmental profile than conventional materials, projects in the UK that incorporate this material tend to perform better in the BREEAM assessment.

5.5.1.5 Green Globes

Green Globes is a comprehensive sustainability rating system that focuses on assessing the environmental performance of buildings. By evaluating commercial buildings for sustainability, health, and resilience, it aims to encourage responsible development practices and enhance the long-term value of properties.

The Green Globes program is unique in that it offers a flexible approach to certification, catering to various types of buildings, from new construction to renovations, interiors, and specialized structures like healthcare facilities. This versatility allows it to be widely applicable across different sectors.

The 1,000-point rating system is a key feature, allowing buildings to score based on a variety of environmental criteria, such as energy efficiency, water use, resource conservation, and overall impact on the surrounding environment. Achieving 35% of the available points is a requirement to earn certification, making it a practical and achievable goal for most projects.

In addition to its environmental focus, Green Globes also helps owners and developers market their sustainability efforts, which can play a crucial role in attracting tenants or customers who prioritize eco-friendly practices. The flexibility of the certification process, which can be tailored to meet specific project needs, further enhances its relevance for a wide range of building types and development goals. This adaptability makes Green Globes an appealing option for projects with diverse sustainability objectives, helping developers highlight their commitment to environmental stewardship while aligning with market demands.

5.5.2 Life Cycle Assessment (LCA)

LCA is a valuable tool used to evaluate the environmental impact of buildings and materials. While there are several definitions of LCA, its core purpose is to analyze the overall environmental burden associated with a building or material throughout its entire life cycle or within a specific phase. There are various stages considered throughout the timber's life cycle. The ISO 14040 series is widely recognized as the standard framework for conducting LCA studies today.

Among the various LCA methodologies, the process-based method is commonly employed in the building construction industry. This method systematically identifies and quantifies the inputs (such as materials and energy resources) and outputs (such as emissions and waste) at each stage of a product's life. Several types of process-based LCA assessments are used, including cradle-to-grave, cradle-to-gate, cradle-to-cradle, and gate-to-gate. Each type provides specific insights relevant to different project timelines and scopes.

By conducting an LCA, stakeholders can gain valuable insights into the environmental performance of buildings. This allows for informed decision-making and helps identify opportunities for improvement. LCA can assess the potential environmental impacts of design choices, material selections, construction techniques, and end-of-life considerations. Such evaluations contribute to the development of more sustainable buildings, fostering resource efficiency, minimizing environmental footprints, and promoting a more holistic approach to building design and construction. Ultimately, LCA is a powerful tool for evaluating and comparing the environmental impacts of building projects, supporting the broader goal of sustainable development in the construction industry.

5.5.2.1 LCA of Mass Timber Buildings

The LCA tool must consider several factors, including labor intensity, equipment requirements, construction speed, and the degree of prefabrication, to provide valuable insights for achieving a more sustainable mass timber building (Crawford & Cadorel, 2017). The steps for conducting an LCA for a mass timber project are similar to those for traditional concrete and steel structures, as outlined below:

Step 1: Goal and Scope Definition

The first step of the LCA involves defining the goal and scope of the assessment, including the project boundaries (e.g., cradle-to-grave, cradle-to-gate). All relevant products and services involved in the project must be identified. Once this is done, the functional unit and system boundaries are specified. The system boundary determines which phases of the life cycle will be considered; for example, focusing only on the construction and use phases would exclude the end-of-life phase from the assessment. The functional unit provides a detailed description of the product or process being analyzed, ensuring that comparisons can be made between alternatives. For example, the functional unit for a car might be the number of miles driven, while for a phone, it could be the number of hours of use. Additionally, impact categories (such as carbon emissions) and data collection requirements are defined (AIA, 2010).

A challenge in using LCA to evaluate MTC is the lack of mass timber buildings that have reached their end-of-life phase. As a result, most LCA studies for mass timber projects focus on a cradle-to-gate scope, with assumptions made for end-of-life phases due to limited data. Consequently, deconstruction, demolition, and waste processing activities are often excluded from these studies (Liang et al., 2021).

Step 2: Inventory Analysis

In this step, an inventory is created that outlines the inputs and outputs, including energy consumption, raw material usage, emissions, and land use. Software tools and databases are essential for ensuring accurate and efficient inventory analysis. One such tool is the Athena Impact Estimator for Buildings, which is commonly used in the North American construction industry to estimate the environmental impact of mass timber projects, especially during later phases (Athena Sustainable Materials Institute, n.d.).

Step 3: Impact Assessment

This step is not necessary if the inventory analysis results are consistent, meaning the products and processes perform well or poorly across all environmental categories. The impact assessment step involves translating the emissions from the inventory analysis (step 2) into specific impacts on human and terrestrial ecosystems, using the predefined impact categories (step 1). While no universally accepted framework exists for this step, the inventory analysis alone is often sufficient to assess the environmental impacts, making the impact assessment step optional (AIA, 2010).

Step 4: Interpretation

In the final step, the results from the previous stages are presented through tables and graphs, helping users evaluate opportunities to reduce environmental impacts. If product or process substitutions are identified, the inventory analysis (step 2) is revisited and the process is repeated. This iterative nature of LCA makes it particularly valuable during the design phase, where adjustments can be made to reduce environmental impacts (AIA, 2010).

Duan et al. (2022) examined 62 peer-reviewed articles on the LCA of MTC. The review found that mass timber buildings generally have lower global warming potential (GWP) and life cycle primary energy (LCPE) compared to reinforced concrete and steel buildings. The same conclusion was arrived at by researchers who conducted a comparative LCA of a 12-story mixed-use building in Portland, Oregon, constructed primarily from mass timber (Liang, et al. 2021). These researchers found that, compared with a similar concrete building, the mass timber building had 18%, 1%, and 47% reduction in global warming, ozone depletion, and eutrophication, respectively.

5.5.3 Environmental Value Engineering (EVE)

To foster a sustainable society, it is essential to make built environment decisions that minimize environmental impact. One effective approach is to evaluate and compare the environmental impact of different building alternatives. EVE is a key methodology that facilitates these comparisons, providing a structured framework for assessing the sustainability of various building options. Developed by Dr. Wilfred H. Roudebush in 1992, EVE integrates Dr. Howard T. Odum's EMERGY analysis with traditional value engineering to offer a comprehensive tool for evaluating the environmental impact of built projects.

EMERGY represents the total energy consumed—both directly and indirectly—through all pathways involved in producing something. The term "EMERGY" combines "energy" and "memory," emphasizing that it accounts not only for direct energy use but also for the accumulated environmental inputs over time. Expressed in Solar EMERGY Joules (SEJ), it quantifies both the energy consumed during production and the environmental cost of those inputs.

Value Engineering (VE), on the other hand, is a structured approach focused on improving the value of a product, project, or process by optimizing its function and cost. The goal is to achieve the required function at the lowest possible cost, without compromising quality, performance, or safety. This process involves analyzing the system's functions, identifying opportunities for cost reduction, and proposing alternatives that maintain or enhance performance.

By combining these two methodologies, EVE allows for the quantification of environmental contributions and impacts, measured in terms of EMERGY, across ten life cycle phases: natural resource formation, exploration and extraction, material production, design, component production, construction, use, demolition, natural resource recycling, and disposal. For each phase, EVE considers the inputs from the environment (E), fuel energy (F), goods (G), and services (S).

EVE is unique because it:

- Uses EMERGY as the primary quantification method, with SEJ as the unit of measurement.
- Considers all relevant inputs, including environmental factors, fuel energy, goods, and services (including labor).
- Encompasses a comprehensive life cycle, consisting of ten distinct phases.
- Can be applied to nearly any situation that involves comparing competing alternatives.

In his analysis of the EMERGY of CLT, Bhattarai (2024) found that CLT had an EMERGY investment ratio of approximately 7.727. This means that for every unit of EMERGY input from the free environment, the process generates about 7.727 units of EMERGY in return. Bhattarai suggests that this reflects the efficiency of CLT in converting environmental inputs into valuable outputs.

5.5.4 Greenhouse Gas Emissions (GHGEs)

GHGEs refer to the release of gases into the Earth's atmosphere that contribute to the greenhouse effect, leading to global warming and climate change. These gases,

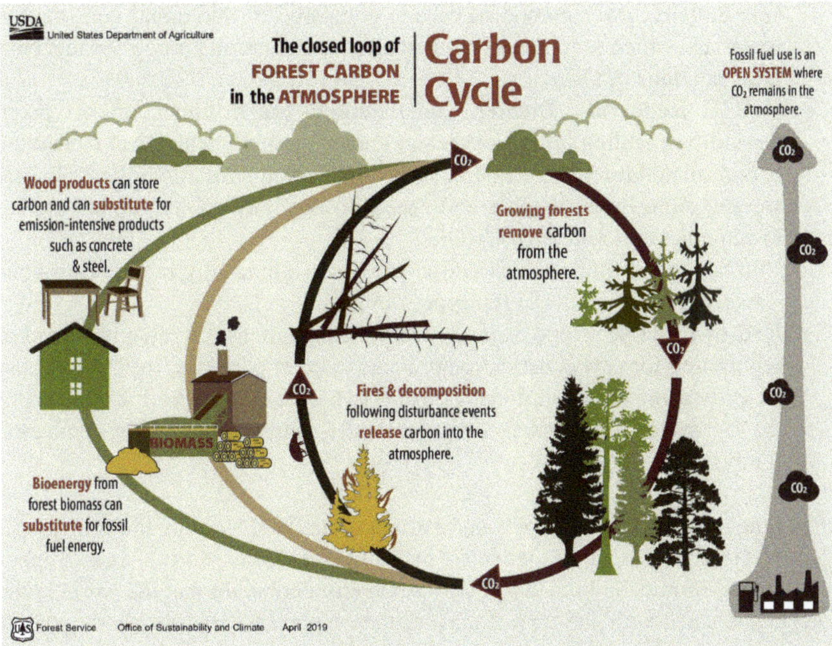

FIGURE 5.5 Various Stages Considered throughout the Timber's Life Cycle (Courtesy of USDA).

which trap heat in the atmosphere, include carbon dioxide (CO_2), methane (CH_4), nitrous oxide (N_2O), and fluorinated gases such as hydrofluorocarbons (HFCs), perfluorocarbons (PFCs), sulfur hexafluoride (SF_6), and nitrogen trifluoride (NF_3).

Mass timber is increasingly recognized for its environmental benefits, particularly in reducing CO_2 across the entire building life cycle, from manufacturing through to end-of-life disposal (see Figure 5.5). Key factors contributing to this reduction include:

- **Carbon Sequestration**: One of the main advantages of mass timber is its ability to act as a carbon sink. As trees grow, they absorb carbon dioxide (CO_2) from the atmosphere, and when the wood is used in construction, the carbon remains stored throughout the building's lifespan. This helps offset emissions from other stages of the building's life cycle. Mass timber continues to store carbon for the duration of the building's existence, effectively keeping the carbon out of the atmosphere.
- **Low Energy Consumption in Manufacturing and Low Embodied Carbon**: While mass timber products require energy for processing, the energy demand is significantly lower than that expended in the production of concrete and steel. As a result, the production of mass timber generates

substantially lower embodied carbon compared to traditional construction materials, which require energy-intensive manufacturing processes that emit large amounts of CO_2.

- **GHGE Reduction During Construction**: Mass timber offers design flexibility, enabling longer spans and greater strength, which can reduce the overall material requirements in some applications. Additionally, its lighter weight, along with the ease and speed of assembly on-site, contributes to lower construction-related GHGEs.

  Prefabricated timber construction can also significantly reduce emissions associated with material transportation.
- **End-of-Life Use**: Wood from deconstructed buildings can often be reused or repurposed for new construction projects or other purposes, further reducing its carbon impact. Additionally, wood is biodegradable, and at the end of its life cycle, it can break down naturally, minimizing its environmental footprint.

One of the most notable examples in Australia is the Forté building in Melbourne's Victoria Harbour District. Completed in 2013, Forté was the first mass timber building in Australia and the tallest wooden apartment building in the world at the time. Constructed with 485 tons of mass timber imported from Europe, the project saved an impressive 1,451 tons of CO_2 emissions when compared to a similar structure made from reinforced concrete. The building also sequestered 761 tons of CO_2 (WoodSolutions, n.d.).

Another landmark mass timber project is the Brock Commons Tallwood House at the University of British Columbia in Vancouver, Canada. This 18-story hybrid building, which combines mass timber floors and columns with a concrete ground floor and elevator cores, is a significant achievement. The timber in the building is estimated to store 1,753 tons of CO_2 and avoid 679 tons of greenhouse gas (GHG) emissions. This reduction is roughly equivalent to removing 510 cars from the road for one year (Williams, 2018).

These examples underscore the potential of mass timber to make a meaningful contribution to reducing carbon emissions in the construction industry while offering other environmental benefits.

5.5.5 Embodied Energy

Energy consumption has become a critical focal point in the global fight against climate change, with buildings playing a significant role due to their embodied energy. Embodied energy, or "embedded energy," represents the total energy expended to produce a material or structure throughout its life cycle. This includes all stages from cradle to gate—spanning material extraction, manufacturing, transportation, and fabrication (Hammond & Jones, 2011).

Mass timber stands out as a highly sustainable alternative, requiring substantially less energy to produce compared to traditional building materials. Furthermore, it demands less energy for transportation and on-site assembly, making it a more efficient and environmentally friendly option for construction. By minimizing both production and operational energy use, mass timber contributes to reducing the overall carbon footprint of building projects.

5.6 MASS TIMBER MANUFACTURING, SUPPLY, AND DISTRIBUTION

The production, supply, and distribution of building materials are crucial factors in achieving a comprehensive and sustainable approach to construction. The construction industry's globalized nature has brought to light the importance of efficient supply chains and distribution systems, especially when it comes to materials like mass timber. Globalization has not only broadened the availability of various materials but has also highlighted the challenges that come with their sourcing, transportation, and timely delivery.

Mass timber, while an environmentally friendly alternative, presents its own set of challenges in terms of supply and distribution. As the demand for mass timber rises, supply chains must be adapted to meet the specific needs of MTC. Unlike more commonly used materials like steel or concrete, mass timber supply is more specialized, requiring a careful balance between sustainable forestry, manufacturing capabilities, and transportation logistics.

Given these constraints, it is crucial that mass timber is considered early in the project development phase. Early engagement with the development team allows for better coordination and ensures that the supply chain for the material can meet the project's timeline. By involving architects, contractors, and suppliers from the outset, projects can account for potential delays and challenges, making the procurement and delivery of mass timber more efficient.

Planning for emerging materials and innovations—such as mass timber—requires that construction professionals consider the entire process, from material sourcing to distribution. This forward-thinking approach will ensure that the materials are available when needed, helping to prevent delays that could arise from supply chain disruptions. Additionally, early consideration of mass timber in the design phase can help identify opportunities to optimize its use, whether through prefabrication, material selection, or logistical considerations.

5.7 MASS TIMBER AND LEAN PRINCIPLES

Lean construction principles have gained widespread acceptance across the construction industry, emphasizing waste reduction and maximizing value for clients. Rooted in lean manufacturing practices, its core principles aim to enhance efficiency, minimize waste, and maximize value throughout construction projects. Project teams that have adopted these methodologies have reported significant improvements in efficiency, productivity, and overall project outcomes. This section highlights key lean principles and examines how they align with MTC.

MTC naturally aligns with lean principles, as both prioritize efficiency, waste reduction, and seamless collaboration. As mentioned before, lean construction emphasizes optimizing production processes, minimizing material waste, and enhancing coordination among stakeholders. These principles support the rapid assembly, precision engineering, and prefabrication advantages inherent in MTC. By reducing project duration, effort, and excess material usage, lean construction helps lower costs associated with planning, budgeting, and implementation.

This section will explore how MTC supports lean construction principles and how its adoption leads to more efficient, sustainable, and cost-effective project delivery.

5.7.1 Lean Construction Principles

According to the Lean Construction Institute (LCI), there are ten key principles:

1. **Specify Value from the Customer's Perspective**: Understand the needs and desires of the customer, delivering value based on these insights.
2. **Identify the Value Stream and Eliminate Waste**: Identify the entire project value stream, from design to delivery, and eliminate activities that do not contribute to the end goal.
3. **Create Flow**: Ensure a smooth, uninterrupted flow of work that minimizes delays and disruptions.
4. **Establish Pull**: Design processes so that materials and resources are only pulled as needed, preventing overproduction and waste.
5. **Seek Perfection**: Continuously improve processes to eliminate waste and strive for perfection.
6. **Emphasize Collaboration and Teamwork**: Foster collaboration among all members of the project team, including owners, architects, contractors, and subcontractors.
7. **Respect People**: Empower workers and promote a culture of respect, trust, and openness.
8. **Optimize the Whole**: Optimize the entire project system, rather than focusing on individual components or tasks.
9. **Deliver Value with the Lowest Possible Cost**: Maximize value while minimizing costs.
10. **Continuously Improve**: Constantly improve processes and practices to enhance efficiency and reduce waste.

5.7.2 Mass Timber and Lean Construction

MTC aligns with lean construction through its focus on off-site manufacturing and modular design. Off-site production ensures high-quality, standardized modular elements are created in controlled environments with efficient operations. These elements are delivered to the construction site, where they can be quickly assembled, minimizing on-site construction time and costs.

MTC offers several advantages, including:

- **Faster On-Site Assembly**: Due to prefabricated components and off-site fabrication, MTC speeds up construction processes.
- **Design for Manufacture and Assembly (DfMA)**: MTC supports DfMA, ensuring that design and production are optimized for lean construction principles (Solarte & Couper, 2022).
- **Lower Delivery Costs**: Mass timber components are lighter than traditional materials like concrete and steel, resulting in lower delivery costs and reduced labor.
- **Ergonomics**: Reduces waste of motion with multiple tools needed to install fasteners in reinforced concrete as opposed to mass timber.

These benefits contribute to reducing costs and waste, improving construction efficiency, and enhancing overall project outcomes.

5.7.3 Impact on Construction Rate and Traffic

MTC results in faster construction rates and less construction traffic. Projects built with mass timber are completed 25% faster than those using concrete, with up to 90% less construction traffic (AWC, n.d.). The efficient use of prefabricated components, off-site fabrication, and DfMA all contribute to reduced disruptions at the construction site.

The prefabricated nature of mass timber components is impactful to project efficiencies. Mass timber is premanufactured in a controlled environment which reduces the likelihood of errors that lead to rework on the jobsite. Similar to structural, the product arrives ready to be installed contributing to construction efficiency during the erection phase. Emphasis is placed on the relationship between the construction and manufacturing team to ensure production and delivery rates are met.

A strong focus on the prefabrication of mass timber components can significantly reduce waste and minimize the need for laydown space on the jobsite. This not only helps reduce congestion but also increases productivity for the erection team and other contractors. Jobsite space is a premium resource, and anything that can free up space for other materials adds value to the construction process.

Not only must materials be procured, but they must also be sequenced properly to avoid inefficiencies. Without careful sequencing, there could be delays or a need to "shake out" materials, which reduces efficiency and can hinder overall project progress. Proper planning ensures that materials arrive and are used at the right time, helping maintain a smooth and efficient workflow.

5.8 CHAPTER SUMMARY

Environmental awareness surrounding the built environment has never been more pronounced. This is evident in the rise of governmental regulations, including carbon taxes, environment, social, and governance (ESG) reporting requirements, and lending guidelines, all of which influence the adoption of sustainable systems and accounting practices. As a result, project teams must possess an in-depth understanding of sustainability systems to navigate these evolving requirements effectively.

Alongside government initiatives, property owners are becoming increasingly well-versed in sustainable construction and building operations. This growing awareness has led to an increase in third-party verified projects, especially those involving mass timber structures, which are still relatively new to the market. Project teams must be prepared for an increased likelihood of additional contractual requirements centered around sustainable systems.

These contractual additions carry far-reaching implications across all aspects of construction management. It is crucial for teams to define and manage these impacts as early and accurately as possible. This calls for a comprehensive management plan that fosters enhanced communication and collaboration among stakeholders. Mass timber structures align closely with sustainable building systems due to their carbon sequestration properties, biophilic benefits, and improved job site efficiency.

Mass timber plays a vital role in all three pillars of sustainability: social, economic, and environmental.

- **Environmental:** Mass timber's primary benefit lies in its low environmental impact. It sequesters carbon, encourages responsible forest management, and contributes to the development of green, energy-efficient buildings. Notably, mass timber buildings are up to twenty-five times more likely to achieve LEED certification, a globally recognized standard for sustainability in building design and construction.
- **Economic:** Mass timber offers a cost-effective alternative to traditional building materials. The reduced need for heavy machinery and labor on-site, combined with the prefabrication of timber components off-site, often results in significant cost savings in both materials and labor. As demand for mass timber continues to grow, investments in improving production processes will likely lower costs and increase material availability in the future.
- **Social:** MTC can enhance the quality of life for building occupants. The natural aesthetic of wood, along with its ability to improve indoor air quality and create a healthier living environment, makes mass timber an attractive choice for both residential and commercial buildings. Furthermore, mass timber can support local economies by creating jobs in forestry, manufacturing, and construction, particularly in regions with abundant forests.

5.9 CHAPTER QUESTIONS

1. The theoretical definition of sustainability in the built environment consists of equal representation of three main stakeholders. These stakeholders include economic, societal, and environmental considerations.
   a. True
   b. False
2. An often overlooked and major benefit in the popularity of mass timber structures is the investment in planting additional trees to expand forest cover. The relative term for this is which of the following:
   a. Deforestation
   b. Afforestation
   c. Reforestation
   d. none the above
3. ______________________refers to the total energy expended to produce a material or structure throughout its life cycle.
   a. Emergy
   b. Embodied carbon
   c. Embodied energy
   d. Carbon footprint
4. Mass timber aligns itself with a circular building economy, much more so than traditional structural steel and concrete.
   a. True
   b. False
5. The integration of more efficient technology in the sawmill industry has translated to raw material yields that almost double 1920s era yields.
   a. True
   b. False
6. Which of the following organizations helps ensure the sourcing of raw lumber is coming from a company that operates the harvesting process sustainably?
   a. FSC
   b. FCS
   c. PFPP
   d. NZE
7. There are five main reforestation techniques, critical for restoring forests and promoting healthy ecosystems.
   a. True
   b. False
8. Mass timber is touted for which of the following sustainability considerations? Choose all that apply:
   a. Third-party sustainability certifications
   b. Carbon sequestration

  c. Renewable resources
  d. Biophilic properties
  e. Reduction of jobsite resources
  f. **All of the above**
  g. A, C, and E
9. How many Lean construction principles are outlined within this chapter?
  a. 5
  b. 7
  c. **10**
  d. 12
10. There are five levels of LEED certification a building can achieve. The levels include Platinum, Gold, Silver, Copper, and Certified.
  a. True
  b. **False**
11. Mass timber projects have a strong correlation of achievement in which of the following third-party certification systems?
  a. **LEED**
  b. NZE
  c. WELL
  d. Green Globes
12. Points achieved in third-party certification systems are often correlated with both an increase in sustainable building attributes and initial cost.
  a. **True**
  b. False
13. Which of the following factors have increased the call for the use of third-party sustainability certifications?
  a. government regulations
  b. consumer awareness
  c. climate change
  d. **all of the above**
  e. none of the above
14. Which third-party sustainable assessment system is considered the "gold" standard in the built environment, when taking into account market utilization?
  a. NZE
  b. WELL
  c. Green Globes
  d. **LEED**
15. Mass timber structures are considered to be carbon emitting versus carbon banks.
  a. True
  b. **False**

REFERENCES

Afgan, N. H., & Carvalho, M. G. (2010). The Knowledge Society: A Sustainability Paradigm. *Cadmus Journal*, 2(1), 1–12.

AIA. (2010). Guide to building life cycle assessment in practice. American Institute of Architects (AIA). https://content.aia.org/sites/default/files/2016-04/Building-Life-Cycle-Assessment-Guide.pdf

American University. (2022). Afforestation and carbon sequestration. Institute for Responsible Carbon Removal. www.american.edu/sis/centers/carbon-removal

Athena Sustainable Materials Institute. (n.d.). Impact estimator for buildings. Retrieved February 26, 2025, from www.athenasmi.org/our-software-data/impact-estimator/

AWC. (n.d.). *Mass timber*. The American Wood Council. https://awc.org/issues/mass-timber/

Bhattarai, A. R. (2024). Analysis of the EMERGY of cross-laminated timber (CLT). *(Master's thesis)*. Kent State University. https://etd.ohiolink.edu/acprod/odb_etd/r/etd/search/10?clear=10&p10_accession_num=kent1714089038602843&session=111909471695920

Callicott, J. B., & Mumford, K. (1997). Ecological Sustainability as a Conservation Concept. *Conservation Biology*, 11(1), 32–40. https://doi.org/10.1046/j.1523-1739.1997.95410.x

Caradonna, J. L. (2014). *Sustainability: A history*. Oxford University Press.

CIB. (1999). *Agenda 21 on sustainable construction*. International Council for Research and Innovation in Building and Construction (CIB).

Crawford, R. H., & Cadorel, X. (2017). A Framework for Assessing the Environmental Benefits of Mass Timber Construction. *Procedia Engineering*, 196, 838–846. https://doi.org/10.1016/j.proeng.2017.08.015

Ding, Y., Liu, X., & Skitmore, M. (2016). Comparative Carbon Footprint Analysis of Steel-Concrete Composite and Timber Buildings. *Journal of Cleaner Production*, 112, 3613–3623.

Duan, Z., Huang, Q., & Zhang, Q. (2022). Life Cycle Assessment of Mass Timber Construction: A Review. *Building and Environment*, 221, 109320. https://doi.org/10.1016/j.buildenv.2022.109320

FAO. (2021). *Sustainable forest management*. Food and Agriculture Organization of the United Nations (FAO). www.fao.org/sustainable-forest-management/en/

Greene, J. (2020, August 12). How much timber does the US harvest, and how is it used?. ResourceWise. www.resourcewise.com/market-watch-blog/how-much-timber-does-the-us-harvest-and-how-is-it-used#:~:text=Per%20recent%20data%20from%20the,the%20US%20economy%20is%20tremendous

Grow Billion Trees. (2025). Afforestation initiatives: Transforming barren lands into thriving forests. https://growbilliontrees.com/pages/afforestation-initiatives-transforming-barren-lands-into-thriving-forests

Hammond, G., & Jones, C. (2011). Embodied carbon: The inventory of carbon and energy. B.S.R.I.A.

Harte, A. (2017). Cross-laminated Timber: Opportunities and Challenges for Urban Building. *Buildings and Cities*, 1(1), 78–89.

Hiziroglu, S. (2017, March 1). Basics of softwood sawmilling – Oklahoma State University. Basics of Softwood Sawmilling | Oklahoma State University. https://extension.okstate.edu/fact-sheets/basics-of-softwood-sawmilling.html#:~:text=In%20the%20mid%2D1920s%2C%20only,value%20from%20each%20log%20processed

IPCC. (2018). Global warming of 1.5°C. An Intergovernmental Panel on Climate Change (IPCC) special report. www.ipcc.ch/sr15/

Kibert, C. J. (2016). Sustainable construction: Green building design and delivery (4th ed.). John Wiley & Sons.

Lehmann, S. (2013). The Carbon Benefits of Sustainably Harvested Wood Products. *Forest Products Journal*, 63(7–8), 275–284.

Liang, S., Gu, H., & Bergman, R. (2021). Environmental Life-Cycle Assessment and Life-cycle Cost Analysis of a High-rise Mass Timber Building: A Case Study in Pacific Northwestern United States. *Sustainability*, 13(14), 7831. https://doi.org/10.3390/su13147831

Mirando, A., Onsarigo, L., & Krsteska, I. (2023). Determining the sustainability of mass timber structures through third-party sustainability assessment systems. *EasyChair*. https://easychair.org/publications/paper/J2mQH

Richard Nixon Museum and Library (2021, April 20). Earth day. www.nixonlibrary.gov/news/earth-day

Pasternack, R., Wishnie, M., Clarke, C., Wang, Y., Belair, E., Marshall, S., Gu, H., Nepal, P., Dolezal, F., Lomax, G., Johnston, C., Felmer, G., Morales-Vera, R., Puettmann, M., & Van den Huevel, R. (2022, January 11). What is the impact of mass timber utilization on climate and forests? MDPI. Retrieved March 13, 2023, from https://www.mdpi.com/2071-1050/14/2/758

Pimm, L. S. (2025, February 4). Deforestation | definition, history, consequences, & facts. Encyclopedia Britannica. www.britannica.com/science/deforestation

Purvis, B., Mao, Y., & Robinson, D. (2019). Three Pillars of Sustainability: In Search of Conceptual Origins. *Sustainability Science*, 14(3), 681–695.

Ramage, M. H., Burridge, H., Buswell, R. A., & Sharma, B. (2017). The Rise of Timber Construction: Opportunities and Challenges. *Royal Society Open Science*, 4(3), 161056.

Schlesinger, P., Powell, B., Silverstein, R., Fishman, T., Esler, J., & Crawford, E. (2021). Global Mass Timber Impact Assessment: A Framework for Assessing Forest and Climate Impacts of Mass Timber. *Forests*, 12(5), 527.

Solarte, A., & Couper, R. (2022). Mass timber buildings: The good, the bad and the ugly. In Marie Frier Hvejsel & Paulo J.S. Cruz (Eds.), *Structures and Architecture. A Viable Urban Perspective? Proceedings of the Fifth International Conference on Structures and Architecture* (pp. 1394–1401). Aalborg, Denmark. https://doi.org/10.1201/9781003023555-166

UNEP. (2019, September 28). Planting trees has never been easier! UN Environment Programme. www.unep.org/news-and-stories/press-release/planting-trees-has-never-been-easier?

U.S. Forest Service. (2022, July 19). Telling the story of America's forests. U.S. Department of Agriculture. Retrieved February 26, 2025, from www.fs.usda.gov/inside-fs/delivering-mission/apply/telling-story-americas-forests

Williams, J. (2018, June 1). Building of the week: Brock commons tallwood house. *The Earthbound Report*. Retrieved February 26, 2025, from https://earthbound.report/2018/06/01/building-of-the-week-brock-commons-tallwood-house/

WoodSolutions. (n.d.). Forte living. Retrieved February 26, 2025, from www.woodsolutions.com.au/case-studies/forte-living

6 Safety and Health

"No Safety, Know Pain. Know Safety, No Pain."—**Author unknown**

LEARNING OBJECTIVES

After studying this chapter, the reader should be able to:

- Understand the importance of safety in construction management
- Understand the safety features of mass timber materials
- Recognize the specific safety considerations for mass timber construction
- Explain the human health factors related to mass timber construction
- Identify the components of a mass timber site-specific safety plan

6.1 SAFETY IN CONSTRUCTION

The health and safety of every stakeholder involved in a construction project is the most critical pillar of construction management. No other factor has as significant an impact on a project's success as safety. While minor delays, unforeseen quality issues, or cost overruns can be managed, major safety incidents or fatalities can derail a project entirely or severely hinder its progress. Therefore, project teams must prioritize safety at every stage of the process. Each project is unique, influenced by various factors, and mass timber projects introduce specific safety considerations that must be carefully addressed. Effective, site-specific safety management is essential for mitigating risks, and this section aims to highlight these unique considerations, offering valuable insights for managing safety on mass timber projects.

Historically, the construction industry has been one of the most hazardous sectors in the United States, contributing to a high number of accidents and fatalities. This has driven substantial investments in accident prevention, including the development of comprehensive policies, the sharing of lessons learned, and an increasing recognition of the link between fiscal success and safety performance.

DOI: 10.1201/9781003482543-6

The Occupational Safety and Health Act of 1970 established a regulatory body responsible for enforcing safety standards, promoting training, and collecting industry data on accidents. In recent years, safety improvements have been spurred by several key factors, resulting in a notable decline in construction-related fatalities (Figure 6.1). A cultural shift within the industry now places a greater emphasis on safety, focusing on worker education, proactive reporting, and continuous improvement. Younger generations, raised in a culture of heightened safety awareness, are increasingly stepping into leadership roles, further reinforcing these values. Additionally, experience modification rates (EMRs) now hold companies accountable by tracking accident rates, which directly impact their fiscal performance. Companies that neglect safety face higher insurance premiums and may even be excluded from bidding on projects.

Technological advancements have also played a crucial role in reducing both the frequency and severity of accidents in construction. Wearable safety devices, material innovations, enhanced machinery safety features, improvements in personal protective equipment (PPE), and the introduction of reporting software all contribute to accident mitigation efforts.

Safety management must begin during the early stages of project development and be integrated throughout every phase of construction. For mass timber projects, teams should consider several factors during the pre-construction phase that may impact safety differently than traditional structures. These factors include ergonomics, environmental exposure, structural integrity, manpower need during the erection phase, and tie-off points for fall protection. Post-construction safety and health benefits, such as biophilic advantages, fire protection properties, and seismic performance, also play an important role in ensuring the ongoing safety

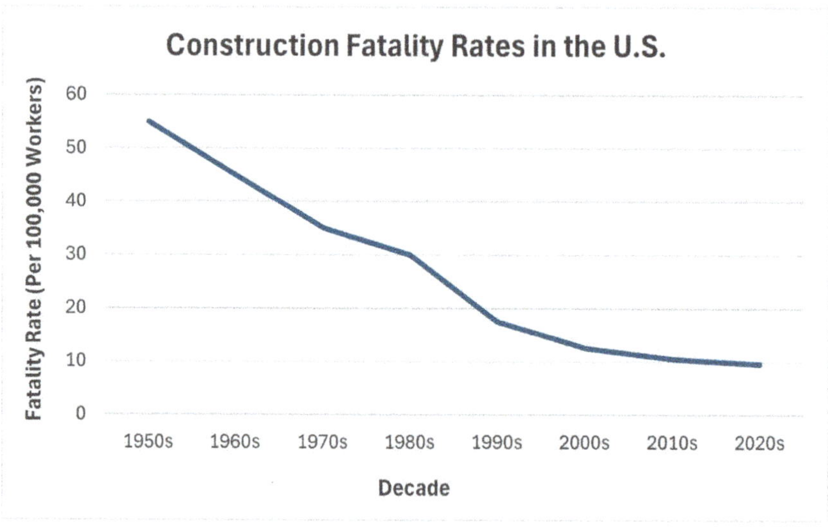

FIGURE 6.1 Construction Fatality Trends in the United States.

of the built environment. Understanding these factors is vital for development, ownership, and construction teams, providing invaluable insight for stakeholders as they make safety-related decisions at each stage of construction.

6.2 DEVELOPING A SITE-SPECIFIC SAFETY PLAN

To successfully deliver and manage any construction project, teams must develop comprehensive, site-specific safety plans. These plans are crucial for identifying and addressing hazards unique to the project. A well-crafted plan typically outlines the scope of work, identifies potential risks, and details the company's safety policies and best practices. The development of site-specific safety plans should begin early in the pre-construction phase and be tailored to account for the unique characteristics of each project. For teams constructing mass timber projects, it is especially important to include information on the erection of mass timber components, ensuring that safety considerations specific to mass timber are thoroughly incorporated.

The responsibility for developing the site-specific safety plan lies with individuals who have a strong understanding of the project requirements, particularly regarding mass timber and its associated safety challenges. While the safety plan may be developed during the proposal phase, at a minimum, it should be shared with the architect and owner during pre-construction, as approval is often required by contract. Throughout the project, the plan can be revised to account for shifting risks, sequencing adjustments, and newly identified hazards.

A well-structured site-specific safety plan typically includes the following components:

- **Scope of Work:** A detailed description of the tasks to be performed and the associated risks involved.
- **Contact Information:** Emergency contacts, project team members, and key stakeholders.
- **Geographic Risks:** Specific hazards related to the job site's location, such as environmental conditions, local regulations, or terrain challenges.
- **Hazard Identification:** A comprehensive list of risks unique to the project, including specific locations and types of potential hazards.
- **Procedures:** Clear safety protocols and practices for managing the identified risks, ensuring that all team members are equipped to respond effectively.
- **Accident Investigation Procedures:** A set of steps to follow in the event of an accident, including reporting, documenting, and analyzing the incident.
- **PPE Guidelines:** Requirements for PPE tailored to the tasks and hazards on the job site.
- **Hazard Communication Plan:** A strategy for effectively communicating hazards to the team, which may include signage, verbal notifications, and regular safety briefings.

It is essential that all stakeholders are familiar with the site-specific safety procedures. This information should be integrated into safety stand-downs, jobsite orientations, and subcontractor documentation requirements to ensure that everyone on site is well-informed and prepared.
Key attributes of a comprehensive safety plan include:

- **Flexibility and Adaptability:** Site conditions can evolve throughout construction, making it essential for the safety plan to remain adaptable to new risks, changes in sequencing, and unexpected circumstances. The plan must be capable of addressing these changes to ensure its ongoing effectiveness throughout the project's lifecycle.
- **Role Clarification:** Clearly defining the roles and responsibilities of each field employee and management team member is critical to maintaining safety on-site. When everyone understands their specific responsibilities, it reduces confusion and improves coordination, leading to a more efficient and safer work environment. Communicating key employee roles, safety processes, and protocols before the erection phase ensures that all team members are prepared and aligned with the safety measures in place.
- **Risk Mitigation Protocols:** The safety plan should include comprehensive guidelines and protocols to proactively manage and mitigate risks. This is particularly important during critical phases of construction, such as the erection of mass timber components.

For mass timber projects, additional safety considerations should be addressed, including:

- **Hot Work on Wood Structures:** Mass timber components require special attention when performing hot work (e.g., welding or cutting) due to fire risk. The safety plan should outline strict protocols for managing hot work activities to prevent fires and ensure the appropriate fire prevention measures are in place.
- **Load Management:** The safety plan must include guidelines for managing the load capacity of mass timber components during installation and throughout the construction process. Given the unique properties of mass timber, it's critical to ensure that each component is properly installed. During construction, live loads from equipment and the associated load distribution should be carefully monitored to ensure safety.
- **Ongoing Monitoring of Structural Erection:** Continuous monitoring during the erection of mass timber structures is necessary to ensure that components are installed safely and according to specifications. This includes checking the integrity of temporary shoring and verifying the stability of components throughout the installation process to prevent accidents and maintain the structure's overall safety.

By addressing these key safety components, the site-specific safety plan can effectively mitigate risks and ensure the health and safety of all stakeholders throughout the construction of mass timber projects.

6.3 CONSTRUCTION PHASE SAFETY CONSIDERATIONS FOR MASS TIMBER

6.3.1 Immediate Structural Integrity

The erection phase is inherently high-risk. Buildings are designed to support their full load only once they are fully erected, squared, and secured. During the erection process, however, the structure is not yet stabilized, creating significant safety hazards. This incomplete state limits the structure's load capacity, restricts access, and often requires additional shoring, which can slow progress and increase risk. Erection teams must remain vigilant in maintaining the structure's stability while minimizing exposure to live loads.

One of the standout advantages of mass timber structures is the immediate structural integrity it provides once installed. While temporary shoring may still be necessary, it is far less extensive than what is required in traditional construction methods. Cross-laminated timber (CLT) panels, when attached to beams and columns, become structurally sound enough to support foot traffic and other small loads almost immediately after installation. This characteristic not only accelerates the construction timeline but also enhances safety during the erection process.

In contrast, cast-in-place concrete requires a curing period before it can fully support loads, during which it is vulnerable to environmental factors and additional loading. This creates potential risks and delays, challenges that mass timber does not face. With mass timber, small equipment, tools, and workers can access areas more quickly and safely, reducing the time spent in potentially hazardous, incomplete stages of construction. This enhanced stability early in the construction process improves both efficiency and safety, offering a clear benefit over traditional building materials.

6.3.2 Less Manpower and Equipment

The availability of qualified manpower is a significant concern in the construction industry. As highlighted in previous chapters, mass timber structures typically require less manpower to complete compared to traditional construction methods. The prefabrication process used in mass timber projects moves much of the workforce from the jobsite to safer, more controlled manufacturing environments. This shift in labor not only enhances efficiency but also offers notable safety benefits during critical phases of construction.

With fewer workers on-site, the risk of injury and safety liabilities during the erection phase is significantly reduced. Fewer workers also lessen the need for staging areas for equipment, porta-johns, and other support facilities, which can pose additional safety hazards. Moreover, the reduced workforce helps minimize congestion and parking issues, making it easier and safer for everyone to access the jobsite. According to the American Wood Council (n.d.), mass timber projects require up to 90% less construction traffic and 75% fewer on-site workers, including erection crews and subcontractors.

This reduction in manpower also results in cleaner job sites. OSHA frequently cites debris, clutter, and poor housekeeping as major safety concerns that contribute to workplace hazards. Since mass timber components are pre-manufactured, there is less waste, leading to a cleaner and safer environment compared to traditional construction methods. This streamlined approach not only improves safety but also enhances overall project efficiency.

6.3.3 Ease of Attaching Tie-Off Points and Handrails

Traditional steel and concrete structures require careful planning and significant effort to install tie-off points and handrails. Whether embedded in concrete or attached to steel, these critical safety features demand extensive work to secure. In contrast, mass timber simplifies the installation of these safety components, allowing for easy placement at multiple points across the structure with minimal effort (see Figure 6.2). This offers greater flexibility for installation teams and reduces limitations when identifying substantial tie-off points. As illustrated in Figure 6.3, there are numerous challenges with installing tie-off anchors in structural steel applications, but such issues do not exist with mass timber.

FIGURE 6.2 Safety Barricades and Tie Off Points.

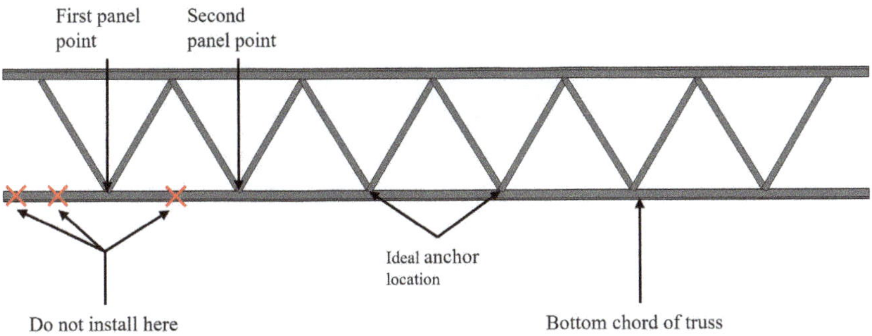

FIGURE 6.3 Limitations of Installation Tie Off Anchors in Structural Steel Applications.

FIGURE 6.4 Harnessing Tie Off Points Directly into the Deck.

Several manufacturers specialize in producing a range of mass timber fall protection products. These manufacturers offer various anchor types, featuring fasteners that can be installed with a simple impact driver (see Figures 6.4 and 6.5). These connectors are rated for fall protection and can be quickly installed across

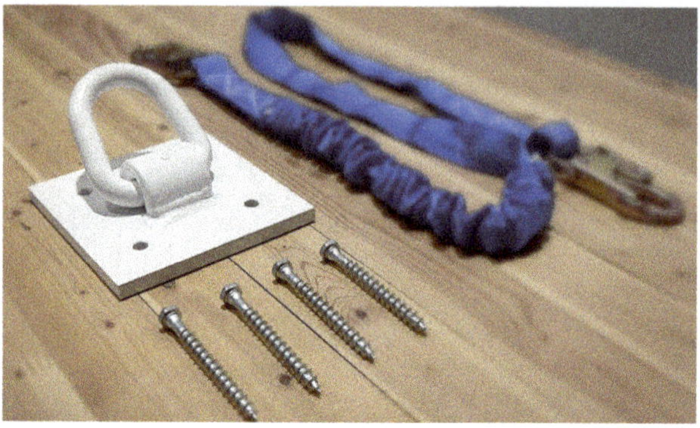

FIGURE 6.5 Tie Off (Courtesy of MTC Solutions).

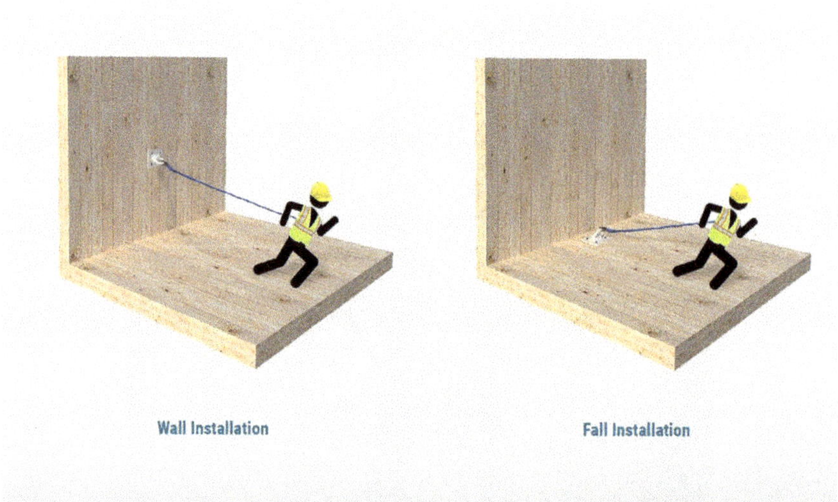

FIGURE 6.6 Tie Off Installation Points (Courtesy of MTC Solutions).

various applications, as shown in Figure 6.6. This enables workers to have greater flexibility and adaptability when managing leading-edge work and fall hazards. By eliminating the need to attach tie-off products to concrete or steel structures, mass timber further enhances worker safety and efficiency, providing a safer, more streamlined approach to fall protection on the jobsite.

6.4 MASS TIMBER SAFETY MISCONCEPTIONS

As the use of mass timber continues to grow in the United States, questions regarding its safety have surfaced. While some of these concerns are valid and merit further investigation, there are also several misconceptions about the safety of mass timber buildings. This section aims to address and clarify some of the most common misunderstandings surrounding mass timber's safety.

6.4.1 Fire

One common misconception about mass timber buildings is that they pose an increased fire risk. Fire safety is a critical consideration in any building design, as inadequate mitigation measures can lead to catastrophic consequences. It is crucial that all buildings, including mass timber structures, are designed to be safe and resilient in the event of a fire. To address this concern, the industry has developed a range of codes and technologies that help mitigate fire risks. These include fire detection systems (such as smoke detectors), fire sprinkler systems, fire alarm systems, fire suppression systems, emergency lighting for safe evacuation, and firestopping and fireproofing materials. These systems are just as essential in mass timber buildings as they are in conventional construction, ensuring that mass timber structures meet comparable fire safety standards.

In addition to fire protection systems, mass timber components—especially when exposed—can be designed to meet fire code regulations. Alternatively, mass timber elements can be concealed with gypsum wallboard, which adds an extra layer of protection that satisfies code requirements and helps prevent damage during a fire (see Figure 6.7). However, covering the mass timber with gypsum

FIGURE 6.7 Drywall Coverage versus Exposed MT.

can negate some of the material's key benefits, particularly those related to the health and well-being of building occupants. While these protective measures are necessary for fire safety, they highlight the balance between ensuring safety and preserving the advantages of mass timber, such as its aesthetic appeal and positive environmental impact. When properly designed and built, mass timber structures offer fire protection that is comparable to, and in some cases exceeds, that of steel and concrete buildings.

6.4.1.1 Exposed Elements (Mass Timber Inherent Fire Resistance)

Mass timber products are known to have superior fire performance when compared to dimensional lumber, largely due to the "charring effect." Fire tests on mass timber elements and assemblies have shown fire resistance ratings of three hours or more, meeting or exceeding code requirements (see Table 6.1 for IBC code requirements). During a fire event, the exposed outer layer of mass timber burns and chars, forming an insulating layer that protects the inner layers from damage as shown in Figure 6.8. This process, known as the "charring effect," not only protects the wood but also releases gases that limit oxygen, further slowing the combustion process. Studies have shown that the charring rates (inches per hour) of these wood elements are predictable. In fact, char rates for heavy timber elements have been

TABLE 6.1
2024 IBC fire-resistance rating requirements for building elements (hours)

Building element	**Type IV**		
	A	**B**	**C**
Primary structural frame[a]	3[b]	2[b]	2[b]
Bearing walls			
Exterior[a]	3	2	2
Interior	3	2	2
Non-bearing walls			
Exterior	Determined based on fire separation distance on the exterior wall and the story in which the wall is located		
Interior	0	0	0
Floor construction and associated secondary structural members	2	2	2
Roof construction and associated secondary structural members	1 ½	1	1

Notes:
[a] Not less than the fire-resistance rating required elsewhere in the code.
[b] Roof supports: fire-resistance ratings of primary structural frame and bearing walls are permitted to be reduced by 1 hour where supporting a roof only.

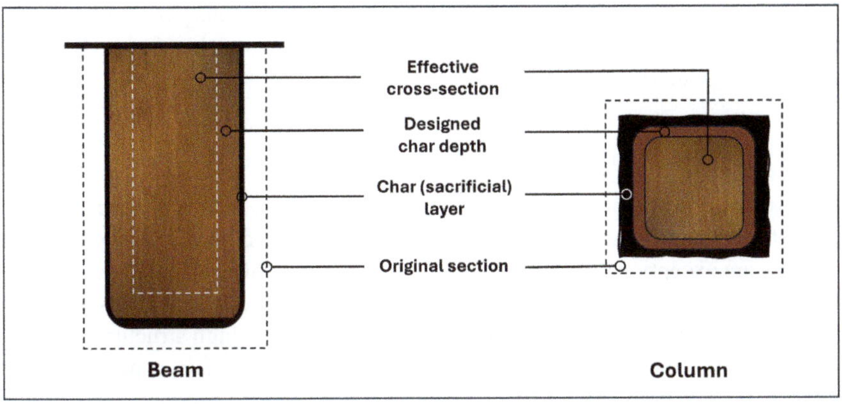

FIGURE 6.8 Mass Timber Design for Fire.

extensively studied and incorporated into building codes and standards for many years. For example, the American Wood Council's "National Design Specification (NDS) for Wood Construction," referenced by the International Building Code (IBC) for calculating fire resistance of exposed wood members, establishes a nominal char rate of approximately 1.5 inches per hour for materials such as sawn lumber, structural glue laminated lumber, laminated veneer lumber, parallel strand lumber, laminated strand lumber, and CLT. Due to its predictable and controlled burning behavior, mass timber can be designed so that reduced cross-sectional dimensions of unburned wood are still sufficient to carry required fire design loads (MyTiCon). This inherent fire resistance makes mass timber a safe and reliable material for fire-resilient structures.

6.4.1.2 Concealed Elements

When required by code, some of the mass timber elements are covered by gypsum wall board which protects the mass timber from damage in a fire event. Gypsum is the key component that serves to curtail the spread of fire within a building. This is because gypsum has chemically combined water that comprises approximately 21% of its weight. When exposed to fire, the water is released slowly in the form of steam through a process called calcination, effectively slowing the spread of fire and transmission of heat. Fire-rated gypsum drywall contains glass fiber reinforcement and other additives that enhance its fire-resistant capabilities. The use of fire-rated gypsum boards combined with the inherent fire resistance of mass timber and fire protection systems help ensure that the structure remains intact longer during a fire, providing critical time for evacuation and firefighting efforts.

Designers must carefully balance the exposure and concealment of mass timber elements to meet both aesthetic and safety objectives. While exposed mass timber enhances the visual appeal of a building and contributes to occupant well-being, it must also be properly protected to meet fire safety standards. Striking the right balance between showcasing the material's natural beauty and ensuring the building's resilience in the event of a fire is essential for achieving both design goals and regulatory compliance.

6.4.2 Structural Performance

Another misconception about mass timber is that it is inherently weaker than concrete and steel. However, this argument is not entirely accurate. While steel and concrete are exceptionally strong in compression, and reinforced concrete and steel perform well under tension, engineered wood products can also withstand both tensile and compressive stresses. Mass timber can achieve high structural strength and durability, comparable to these traditional building materials, especially when properly designed and engineered. As an engineered wood product, mass timber mitigates the effects of natural wood defects, resulting in predictable and reliable strength characteristics. Additionally, the structural components of a mass timber building weigh approximately 20% of a comparable concrete structure (University of Toronto, 2025), which, as previously mentioned, results in smaller and more efficient foundations.

6.4.3 Shock and Seismic Resilience

Closely related to the misconception about mass timber's strength is its ability to withstand shock and seismic activity. Due to its flexibility and energy-absorbing qualities, mass timber can perform well in earthquake-prone areas when properly designed. This is primarily due to the following reasons:

1. **Mass Timber's High Strength-to-Weight Ratio**: Mass timber buildings are lighter than concrete structures, which helps reduce the risk of failure during earthquakes. The high strength-to-weight ratio of mass timber allows it to perform effectively in seismic conditions. This is primarily due to the significant reduction in inertial seismic forces, as the lighter mass of the building generates less force during movement, improving overall stability during an earthquake.
2. **Ductility of Wood in Seismic Events**: Wood is a naturally ductile material, meaning it has the ability to flex and return to its original shape without permanent damage. This inherent flexibility allows mass timber to perform exceptionally well when subjected to seismic forces, as it can absorb and dissipate energy without failing, enhancing the building's overall resilience during earthquakes.

In May 2023, seismic tests were conducted on a 10-story mass timber building project (Figure 6.9) at the Englekirk Structural Engineering Center at the University of California, San Diego (UC San Diego). The structure was built specifically for testing as part of the Natural Hazards Engineering Research Infrastructure (NHERI)

FIGURE 6.9 NHERI TallWood Seismic Test Project (Courtesy of the U.S. National Science Foundation).

TallWood Project, an industry-wide initiative exploring the seismic resilience of mass timber. A variety of mass timber materials were incorporated into the design, including CLT, dowel-laminated timber (DLT), glue-laminated timber (GLT), nail-laminated timber (NLT), and structural composite lumber (SCL) Figure 6.10. The tests revealed that mass timber structures can withstand the forces of extreme seismic activity.

6.5 HUMAN HEALTH FACTORS

6.5.1 Exposure to Elements

Recent safety data highlight a concerning increase in worker illnesses due to exposure to various hazards on construction jobsites, with silica exposure emerging as a major concern. Silica dust can result from activities such as drilling, cutting, and mixing cement products, and prolonged exposure may lead to silicosis, an

FIGURE 6.10 Different MT Types on NHERI TallWood Project (Courtesy of Freres Engineered Wood).

untreatable lung condition. In response to this growing risk, OSHA updated its guidelines in 2016, lowering the permissible exposure limit (PEL) for respirable crystalline silica from 250 micrograms per cubic meter of air over an 8-hour shift to 50 micrograms per cubic meter (Occupational Safety and Health Administration, n.d.). A further update in 2020–2021 mandated that employers comply with these new standards, and required medical surveillance for employees exposed at or above the action level for 30 or more days. Managing silica exposure is a critical consideration for project teams moving forward.

Because mass timber projects use significantly less concrete compared to traditional cast-in-place concrete projects, the potential for silica exposure is notably lower. Concrete foundations are also typically smaller in mass timber buildings, and trades rarely need to penetrate concrete for tasks like fastening equipment, ductwork, conduit, or cabling. Instead, they can screw directly into solid wood, eliminating one of the main sources of silica exposure—dust generated during drilling and cutting concrete with hand tools (see Figure 6.11).

FIGURE 6.11 Excessive Amounts of Silica Generated onsite from Cutting Concrete (Courtesy of Texas.gov).

Silica exposure is also a concern in the drywall industry, where finishing and sanding surfaces releases airborne particles that can irritate the eyes, nose, throat, and respiratory system. In mass timber structures, drywall usage is minimized to expose as much timber as possible, further reducing the potential for silica dust exposure. For example, the INTRO Cleveland project covered only 48% of its interior surfaces with drywall. Similar projects have reported significant reductions in drywall usage, which not only lowers project costs but also reduces exposure to harmful substances. Additionally, less drywall means fewer surfaces that require painting, further reducing exposure to harmful respiratory elements found in paints and finishes. By incorporating mass timber, project teams decrease the amount of respirable silica dust and aerosols, benefiting both installation trades and other workers on site.

Another advantage of mass timber projects is the reduction in noise during the installation process. With less reliance on heavy machinery and larger power tools—and with wood naturally absorbing sound—mass timber installations produce less noise and fewer distractions. In contrast, cast-in-place concrete projects typically require loud pump trucks, heavy equipment, and larger teams. Reducing noise levels not only minimizes distractions on the job site but also has a positive impact on long-term hearing health. By using mass timber, projects contribute to a safer, healthier environment for workers in both the short and long term (Brind'Amour & Bertrand, 2023).

6.5.2 Ergonomics

Ergonomic concerns have long been a significant issue in construction management, particularly for trade employees. Both long-term health impacts and short-term efficiency are influenced by the ergonomics of each trade. Ergonomics refers to the study of how individuals perform tasks within their work environment, with a focus on maximizing efficiency and minimizing physical strain. In construction, determining efficiency can be challenging due to the diverse range of movements and tasks required for different trades. For example, a plumber might spend one day installing 4″ PVC pipes in a trench, and the next day soldering overhead copper lines—each task requiring different movements, materials, tools, and layouts. Additionally, the methods used by trades to complete their work significantly impact both performance and health. Workers exposed to certain installation techniques often develop long-term health issues, such as floor setters who spend long hours kneeling, which can lead to lower back and knee problems. By offering solutions to improve installation efficiency, project teams can also help reduce the long-term health impacts on workers.

Mass timber's inherent properties not only enhance the structural installation process but also have a profound positive impact on the subsequent trades. Mechanical, electrical, low voltage, fire protection, and other trades that fasten materials to the structure benefit from a less physically demanding installation process. Simply put, drilling into wood is far easier and less physically taxing than drilling into steel or concrete. As discussed in the scheduling section, efficiency gains for MEPFP trades are directly linked to fewer movements, as the need for pre-drilling fastener holes is eliminated. Fewer movements not only improve efficiency but also reduce the risk of injury.

Communicating the benefits of increased efficiency and reduced physical strain to subcontractors before the bidding process is essential, as it can significantly impact project schedules, costs, and trade personnel selection. Tradespeople working on mass timber projects experience less bodily strain, leading to quicker installation times and, ultimately, a safer, more productive work environment.

6.5.3 Post-Occupancy Health and Biophilic Design in Mass Timber Projects

Post-occupancy health is a crucial consideration in the development of mass timber projects. Since humans spend approximately 90% of their time indoors, the quality, nature, and interior elements of the spaces we inhabit have a profound impact on our overall well-being (Environmental Protection Agency, 2024). As highlighted throughout this book, project teams should have a thorough understanding of the health benefits associated with different building materials, as these factors often play a key role in the decision-making process for structural systems. One particularly impactful approach is biophilic design, an innovative methodology focused on enhancing, maintaining, and restoring natural experiences within the built environment.

Numerous studies have shown that biophilic design can lead to significant improvements in health outcomes, especially in mass timber projects. The natural aesthetic qualities of wood, combined with its tactile and sensory appeal, contribute to reduced stress levels, increased cognitive function, and overall better mental health. By integrating elements of nature—such as natural light, plant life, and wood finishes—into the built environment, mass timber buildings not only foster a connection to the outdoors but also create spaces that promote long-term health and well-being for occupants.

6.5.3.1 Patterns of Biophilic Design

Biophilic design can be implemented through a variety of patterns that strengthen the connection between the built environment and nature. These patterns can be categorized into three key groups, each focusing on different aspects of integrating natural elements into architectural design:

1. **Nature in the Space Patterns:**
    - **Presence of Water**: Incorporating water features such as fountains, streams, or indoor water walls that engage the senses and contribute to a calming atmosphere.
    - **Dynamic and Diffuse Light**: Allowing natural light to flow into the space in varied ways, such as through skylights, large windows, or light wells, which change throughout the day to create a dynamic, ever-evolving environment.
    - **Visual Connection with Nature**: Ensuring that occupants have views of the natural environment outside, whether it's through large windows, green walls, or outdoor gardens, to foster a sense of connection to the outdoors.
2. **Natural Analogues Patterns:**
    - **Biomorphic Forms and Patterns**: Using organic shapes, textures, and patterns that mimic forms found in nature, such as leaf-like patterns, flowing curves, or cellular structures that evoke the natural world.
    - **Material Connection with Nature**: Incorporating natural materials like wood, stone, and plant-based materials that connect the interior environment to the earth and contribute to a sense of warmth and authenticity.
3. **Nature of the Space Patterns:**
    - **Prospect (Wide Views)**: Designing spaces that offer expansive views, creating a sense of openness and connection to the broader environment, allowing occupants to see and appreciate the surrounding landscape.
    - **Refuge (Areas for Retreat)**: Providing spaces that offer privacy, shelter, and a sense of safety, such as alcoves, lounges, or quiet corners, where individuals can retreat and recharge.
    - **Mystery (Intrigue and Discovery)**: Incorporating elements that spark curiosity, such as hidden passageways, layered visual experiences, or intriguing architectural details, encouraging exploration and surprise.
    - **Risk/Peril (Controlled Elements that Engage the Senses)**: Introducing elements that evoke controlled, safe risk or excitement, such as the use

of dramatic lighting, textures, or height variations that engage the senses and stimulate a sense of adventure.

By incorporating these biophilic design patterns, project teams can create environments that promote occupant health, well-being, and productivity, while simultaneously enhancing the aesthetic appeal and overall user experience of mass timber structures.

6.5.3.2 Biophilic Properties of Mass Timber

The biophilic properties of mass timber offer numerous health benefits for occupants, many of which contribute to a higher quality of life. Research supports the following positive effects of incorporating mass timber into the built environment:

1. **Stress Reduction**: Exposure to wood has been shown to reduce stress levels and improve overall mood, creating a more relaxing environment.
2. **Improved Productivity**: Workspaces with exposed wood elements have demonstrated increased productivity, as the presence of natural materials helps enhance focus and concentration.
3. **Lower Blood Pressure**: Studies indicate a direct link between exposure to wood and lower blood pressure, contributing to overall cardiovascular health.
4. **Connection to Nature**: Occupants report feeling a stronger connection to nature when surrounded by wood elements, fostering a sense of calm and well-being that extends to various aspects of their health.
5. **Thermal Comfort**: Wood's low thermal conductivity makes it feel warmer to the touch compared to materials like concrete or steel, enhancing occupant comfort, particularly in colder climates.
6. **Aesthetic Appeal**: The natural, warm appearance of wood improves the visual experience of a space, creating a more inviting and aesthetically pleasing environment.

Exposure to nature and natural elements, such as wood, has been shown to reduce stress, enhance cognitive function, improve attention, and boost psychological well-being (Grinde & Patil, 2009; Yin et al., 2018). The biophilic properties of timber not only contribute to better health outcomes but also enhance productivity. This can translate into significant cost benefits, particularly for employers. For example, a company that invests in mass timber spaces may see fewer sick days and higher employee productivity, ultimately benefiting its bottom line.

Additionally, the health and productivity advantages of biophilic design have been linked to increased rental rates and higher asking prices for mass timber buildings. These positive impacts demonstrate that investing in biophilic elements not only enhances occupant well-being but also proves financially beneficial for building owners. As such, these benefits should be considered early in the development process, helping to inform key decisions for stakeholders and ensuring the long-term success of mass timber projects.

6.5.3.3 Incorporating Biophilic Properties into Mass Timber Projects

To fully leverage the benefits of biophilic design, project teams can incorporate a range of specific features into mass timber projects including:

1. **Use of Natural Materials (Exposed Mass Timber):**
    - Make the mass timber structure—such as beams, columns, and ceilings—visible within the design. This not only highlights the inherent beauty of the wood but also fosters a sense of connection to nature and the natural world. Exposed timber helps create an inviting, warm atmosphere that positively impacts occupant comfort and mood.
2. **Maximize Natural Light:**
    - Incorporate large windows, skylights, and atriums to allow an abundance of natural light to enter the space. Careful consideration of shading devices or design solutions can reduce reliance on artificial lighting, making the space more energy-efficient while enhancing occupants' exposure to daylight, which has been linked to improved mood, health, and productivity.
3. **Integrate Views of Nature:**
    - Position the building to optimize views of the surrounding natural environment. This can be achieved through thoughtful site planning and the strategic placement of windows and outdoor spaces that connect the indoors to the outdoors. Additionally, landscaping around the building can amplify the visual connection to nature and provide spaces for relaxation or reflection.
4. **Incorporate Vegetation:**
    - Bring nature indoors by integrating green walls, vertical gardens, planters, and even trees. This not only improves air quality but also provides aesthetic and mental health benefits. Rooftop gardens can offer additional ecological benefits and provide an urban oasis for building occupants, helping to reduce stress and promote well-being.
5. **Design for Sensory Variety:**
    - Create spaces that engage the senses through a variety of design elements. For example:
        - **Visual**: Use organic, nature-inspired forms or patterns that mimic the natural world.
        - **Tactile**: Incorporate natural textures (e.g., wood, stone, or fabric) that offer pleasant tactile experiences.
        - **Auditory**: Integrate sound-reducing materials or design elements to minimize noise pollution and enhance auditory comfort.

- **Olfactory**: Highlight the natural scent of timber or integrate elements like scented plants to provide calming fragrances, promoting a multisensory connection to nature.

6. **Foster Community and Interaction:**
   - Design communal spaces that encourage social interaction and a sense of belonging. These spaces can include lounges, shared workspaces, or gathering areas that promote connectivity among building occupants. Creating a sense of community is an essential part of biophilic design, fostering a positive and supportive environment for both personal well-being and professional collaboration.

By incorporating these biophilic design principles into mass timber projects, teams can create healthier, more productive environments that benefit both occupants and owners. Not only do these features improve mental and physical health, but they also enhance the building's financial performance. As occupants experience reduced stress, improved focus, and better overall well-being, owners benefit from greater tenant satisfaction, higher rental rates, and increased market value.

6.6 CHAPTER SUMMARY

Safety is the most critical pillar of construction. Organizations that prioritize safety not only protect their teams but also enjoy significant fiscal benefits. The health and safety of both construction workers and building occupants should always be the top priority. To meet safety standards, project teams must develop comprehensive, site-specific safety plans that are implemented and managed throughout the construction process.

In addition to managing site operations, owners and designers must proactively consider the health and safety of the building's future occupants. Early involvement from all stakeholders during the design and pre-construction phases is crucial to identifying and addressing potential risks. Teams that integrate safety strategies throughout the project lifecycle are more likely to achieve greater success. Investing in safety reduces risks, boosts productivity, and improves working conditions—ultimately supporting project success.

Mass timber structures offer specific benefits that contribute to safer, healthier building operations. This chapter highlights key advantages of mass timber in terms of safety, including:

1. **Environmental Benefits**: Mass timber has a lower environmental impact compared to traditional construction materials, contributing to safer and healthier environments for both workers and occupants.
2. **Biophilic Properties**: The natural aesthetics and material properties of mass timber, such as warmth and visual appeal, promote well-being and reduce stress for both construction workers and building occupants.

3. **Cleaner Job Sites**: Mass timber projects typically generate less waste and require fewer resources on-site, resulting in cleaner, more organized job sites that reduce safety hazards.
4. **Flexibility in Fall Protection**: The structural nature of mass timber offers versatile fall protection strategies, enhancing safety during the erection phase and beyond.

By leveraging these advantages, construction teams can ensure that mass timber structures are built with optimal safety, contributing to both project efficiency and the well-being of all involved.

6.7 CHAPTER QUESTIONS

1. Mass timber structures produce less airborne silica dust than CIP and steel structures?
   a. True
   b. False
2. Choose one of the main operational safety benefits of mass timber structures:
   a. Reduced silica dust
   b. Decreased amount of drywall
   c. Smaller foundations
   d. Increased availability of tie-off points
3. Mass timber projects require less manpower to erect the structure when compared to cast-in-place and steel structures.
   a. True
   b. False
4. Mass timber products are known to have superior fire performance compared to dimensional lumber due to which of the following?
   a. sealers
   b. charring effect
   c. relationship to sprinkler systems
   d. fire retardant adhesives
5. Projects that utilize mass timber structures often require more manpower to install both the structure and subsequent trades work.
   a. True
   b. False
6. A major contributor to the reduction of injuries and deaths in the construction industry, the Occupational Safety and Health Act was passed by the United States congress in what year?
   a. 1969
   b. 1923
   c. 1987
   d. 1970

7. There are three distinct patterns outlined in biophilic properties; these include (choose all that apply):
   a. **Nature in the space patterns**
   b. **Nature of the space patterns**
   c. Biological threshold patterns
   d. **Natural analogue patterns**
   e. None of the above
8. Mass timber is known for its positive effect on ________, which focuses on how employees perform their tasks.
   a. efficacy
   b. **ergonomics**
   c. sequencing
   d. morale
9. Mass timber is said to be quieter when compared to traditional construction methodologies, positively affecting jobsite communication and limiting inner ear damage.
   a. **True**
   b. False
10. An increase in the number of _____ ______ points offers construction workers more flexibility in fall protection options.
    a. structural elements
    b. leading edge
    c. **tie-off**
    d. slip protection
11. Exposed wood is said to be a contributor to the reduction of stress of occupants.
    a. **True**
    b. False
12. There are three key components to a site-specific safety plan. Choose all that apply:
    a. **Flexibility and Adaptability**
    b. **Role clarification**
    c. Cost-effectiveness
    d. **Risk Mitigation protocols**
    e. All of the above
    f. None of the above
13. What two mass timber properties enable it to perform effectively in shock and seismic events?
    a. **High strength-to-weight ratio**
    b. Fire resistance
    c. **Ductility**
    d. Thickness

14. Mass timber installation positively impacts the ergonomics of structural installers, as well as subsequent trades such as mechanical, electrical, and fire suppression.
    a. **True**
    b. False
15. What are the effects of mass timber's biophilic properties?
    a. **Stress reduction**
    b. Lower blood sugar
    c. **Enhanced thermal comfort**
    d. Improved vision

REFERENCES

American Wood Council. (n.d.). Mass timber. https://awc.org/issues/mass-timber/

Brind'Amour, M., & Bertrand, S. (2023, September 29). *Fact sheet: Building sustainably: Mass Timber (September 2023)*. EESI. www.eesi.org/papers/view/fact-sheet-building-sustainably-mass-timber-september-2023#:~:text=Mass%20timber%20building%20construction%20can,Workforce%20Growth

Grinde, B., & Patil, G. G. (2009). Biophilia: Does Visual Contact with Nature Impact on Health and Well-Being? *International Journal of Environmental Research and Public Health*, 6(9), 2332–2343.

Environmental Protection Agency. (2024, July 8). Indoor air quality. EPA. www.epa.gov/report-environment/indoor-air-quality#:~:text=Importance%20of%20Indoor%20Air%20Quality,-%E2%80%9CIndoor%20air%20quality&text=Americans%2C%20on%20average%2C%20spend%20approximately,higher%20than%20typical%20outdoor%20concentrations

Occupational Safety and Health Administration. (n.d.). 1910.1053 – Respirable crystalline silica. www.osha.gov/laws-regs/regulations/standardnumber/1910/1910.1053

University of Toronto . (2025). *FAQ*. Mass Timber Institute. https://academic.daniels.utoronto.ca/masstimberinstitute/faq/#:~:text=Mass%20timber's%20light%20weight%20lends%20to%20its,its%20weight%20enables%20outstanding%20performance%20in%20earthquakes

Yin, J., Zhu, S., MacNaughton, P., Allen, J. G., & Spengler, J. D. (2018). Physiological and Cognitive Performance of Exposure to Biophilic Indoor Environment. *Building and Environment*, 132, 255–262.

Case Studies

To ensure consistency and ease of understanding, the case study formats will align with the structure of the book chapters. Each case study begins with an introductory overview of the project, followed by detailed discussions on scheduling, estimating, quality control, sustainability, and safety. Visuals, examples, and relevant data will accompany each section to provide a comprehensive view. Special emphasis will be placed on key pillars of construction management, with each case study expanding on the most significant aspects that add value to the reader.

Each case study has been thoughtfully chosen to represent the wide-ranging applications and geographical diversity of mass timber. All examples focus on large-scale mass timber projects, each presenting a distinct set of challenges and opportunities, offering rich insights to the reader. This collection of projects spans a variety of types and scales, providing a well-rounded perspective on the material's versatility and potential.

The cases are listed in the order below:

1. INTRO: Cleveland, Ohio. This 512,000 SF multi-use commercial structure is, at the time of this publication, the largest mass timber structure in the United States.
2. Ascent: Milwaukee, Wisconsin. This 25-story mass timber tower is the tallest mass timber building in the world, showcasing the potential for mass timber to be used as structural elements on tall buildings.
3. T3 Tower: Minneapolis, Minnesota. One of the first large scale mass timber projects, this site broke barriers and records alike. This seven-story building set precedent for mass timber construction in the commercial sector.
4. Spring Street: Los Angeles, California. Erected on existing foundations, this project offers new office spaces for businesses, highlighting the potential for mass timber to enhance existing buildings and increase capacity.
5. The High Line—Moynihan Connector Timber Bridge: New York City, New York. A groundbreaking pedestrian bridge project, comprising two distinct bridges incorporating the use of concrete, steel and mass timber to enhance connectivity and sustainability.
6. Walmart's Home Office: Bentonville, Arkansas. Walmart's world headquarters campus, consisting of multiple, interconnected structures including, gyms, office spaces, daycare centers, and more.

CASE STUDY I : INTRO CLEVELAND, CLEVELAND, OH

The city of Cleveland, OH, is known for many things, but being home to record-breaking buildings has not traditionally been one of them. However, in 2019, Harbor Bay Real Estate Advisors, a cutting-edge developer based in Northbrook, Illinois, approached the city with a project that would forever change its skyline and the construction industry. At that time, large-scale mass timber projects were beginning to emerge in major cities across the country, but nothing of this size, cost, and impact had been attempted before.

This ambitious project required collaboration between the city, suppliers, developers, contractors, and a wide range of other stakeholders. As a groundbreaking initiative, numerous challenges arose during construction, each demanding innovative solutions. This case study not only highlights the overall project but also delves into specific challenges and their resolutions, aligned with the core principles of construction management. Throughout the book, several sections reference this case study to illustrate key concepts and strategies.

As one of the pioneering large-scale mass timber projects, INTRO Cleveland has generated a wealth of interviews, project documentation, case studies, articles, and more, all of which readers are encouraged to explore alongside the insights provided in this case study. Given the depth and breadth of the project, it would be impossible to cover every important aspect here, which is why this case study focuses on select key pillars that are most relevant to the discussion.

The authors dedicated significant time and effort to this project, engaging with subcontractors, developers, and the construction team on a weekly basis for two years. This hands-on involvement, combined with a vast amount of collected data and personal experiences, is reflected throughout the case study. The experience gained from working on INTRO Cleveland serves as the foundation for this text, offering readers numerous valuable lessons drawn from first hand involvement in the project.

PROJECT DETAILS

When completed in 2022, INTRO Cleveland became the tallest and largest mass timber structure in the United States. Located in Cleveland, Ohio, this mixed-use residential development stands nine stories tall and spans 512,000 square feet (HP Architecture, n.d.). At a height of 115 feet, the building houses 35,000 square feet of retail space, 12,000 square feet of event space, and 298 residential units. The mass timber frame is supported by a cast-in-place concrete foundation, which also includes 342 parking spaces. In addition to its residential and commercial offerings, the building features a 60-foot long rooftop pool, lounge areas, one acre of public green space, a fitness center, and a dog park. The total construction cost was \$145 million, creating over 730 jobs for the city. The project also achieved LEED Gold Status from the U.S. Green Building Council (USGBC). The innovative hybrid structural system combines cross-laminated timber (CLT) and glulam beams with a reinforced concrete podium at the base.

INTRO Cleveland is located at the intersection of West 25th Street and Lorain Avenue in Cleveland's Ohio City neighborhood. The project is owned by Harbor Bay Developers, with Hartshorne Plunkard Architects serving as the architectural firm and Panzica Construction as the general contractor. Structural engineering was provided by Forefront Structural Engineers, with Fast+Epp handling the extensive mass timber engineering. To complete the project, 5,650 m^3 of CLT and 1,500 m^3 of glulam sourced from sustainable forests were fabricated and supplied by Binderholz, an Austria-based mass timber manufacturer (Binderholz, n.d.).

As the largest structure to utilize mass timber, one of the first challenges for the development team was securing approval from the city of Cleveland. Local municipalities are typically cautious when it comes to large-scale projects using non-traditional construction methods. This required Harbor Bay to assemble a team of lawyers, engineers, and specification consultants to engage with the city for building permit approvals. After months of negotiations and providing substantial evidence to support the project's feasibility, the city granted the necessary building permits. However, specific negotiations would continue throughout construction. For instance, drywall coverage percentages were a point of contention even during installation. The building code at the time allowed for 80% of the ceiling to be covered with drywall, while the INTRO team requested a significantly lower percentage. Acoustic performance was another area of debate, requiring ongoing resolution as the project progressed.

To navigate the complexities of this unprecedented project, Harbor Bay enlisted a comprehensive team of professionals. Admittedly, because this was a first for everyone involved, the initial team was larger than typically required. The core team for the mass timber structural division included the following:

- **Structural Engineers**: Fast+Epp and Forefront Structural Engineers, Inc.
- **Architect**: Hartshorne Plunkard Architecture
- **Mass Timber Hardware Supplier**: Rothoblass
- **Timber Procurement and Installation**: Seagate Mass Timber, Inc.
- **Additional Installation Hardware**: Simpson Strong-Tie

During the contract negotiations, discussions around project design and material procurement were also underway. The team ultimately chose Binderholz as the mass timber supplier due to their competitive pricing, design support, and solid industry reputation. Procuring the materials, however, presented its own set of challenges.

International procurement of commodity-based materials often involves logistical hurdles, and importing mass timber from Austria to Cleveland was no exception. To mitigate this risk, Harbor Bay forged a partnership with the Port of Cleveland, which provided both financial and logistical assistance. The project team began discussions with local professionals in September 2019, which presented two major challenges. The first was navigating the financial options available through the Port, while the second was figuring out how to bring mass timber directly from Austria to Cleveland.

The Port of Cleveland has a long history of financing capital projects, helping fund 150 projects totaling $4.5 billion since 1993. The Port offers financing in the form of taxable lease revenue bonds, providing tax advantages to developers. Cleveland is one of five U.S. ports that has its own internal bond fund, which can be used for real estate and development activities up to $10 million. However, understanding the parameters of the program and ensuring compliance required several rounds of negotiations. After multiple meetings, the development team and the Port reached an agreement on the financial terms. Harbor Bay has credited this relationship with making the project possible, noting that without the Port's financial and logistical support, INTRO Cleveland might not have been realized.

The second challenge was the maritime logistics of importing the mass timber from Austria. Shipping, especially during the COVID-19 pandemic, posed significant challenges. The Port of Cleveland had never before financed or provided logistical support for such a project, marking another first for the relationship. The key challenge was getting an international vessel through the Cleveland-Europe Express (CEE) service, which offers competitive transit times for cargo owners. The CEE is the only container service on the Great Lakes that handles both imports and exports, with an average sailing time of 13 days between Cleveland and Antwerp.

The logistical process began with assessing the quantity and timing of materials, as well as evaluating storage and transportation options from the port to the project site. A major advantage was the project's proximity to the Port of Cleveland—just under two miles away. Additionally, the port offered 320,000 square feet of storage space to accommodate the timing of deliveries.

The mass timber components were prepared for shipping in Fügen, Austria, and loaded onto a Norwegian vessel through the Port of Cleveland's partner, the Spliethoff Group. The vessel, capable of carrying up to 20,000 tons of cargo, embarked on a 16-day transatlantic journey, passing through the St. Lawrence waterway and arriving at the Port of Cleveland. The total duration from shipment to arrival averaged 22 days. Upon arrival, the materials were transported by truck to the project site for installation. Prior to distribution to the jobsite, metal connections were installed on the glulam beams and columns to ensure they would be ready for installation upon arrival.

This extensive process required significant planning and resources from the development team. When importing materials from overseas manufacturers, construction teams must carefully plan for the logistics involved. In this case, the project team successfully addressed both financing and logistical challenges with one partner—the Port of Cleveland. However, other geographic locations may present different challenges. Therefore, it's crucial to account for upfront planning, as well as pre-construction costs, in both estimates and project timelines.

SCHEDULING

Harbor Bay embraced the motto "Non-Stop to the Top," which was prominently festooned across jobsite trailers, meeting boards, office walls, and even email signatures. Established by the development team, this creed reflected their ambition to erect the structure with such efficiency that work would progress seamlessly through building enclosure. The team aimed to fully leverage the erection efficiency of mass timber. Harbor Bay's Vice President of Construction, James Litwin, described the process as a predictable building experience, emphasizing efficient movement through each phase of construction (CBCTC, 2020). However, while inspiring, the motto proved challenging to achieve in practice. Despite extensive pre-planning, maintaining uninterrupted, efficient installation remained a complex task. This section explores the overall construction duration, key statistics, innovative solutions, and lessons learned related to the project's schedule.

The first major challenge the team faced was the lack of experience among most of the construction crew in mass timber construction. The only contractor on-site with prior expertise was Seagate Mass Timber Inc., which provided field installation specialists to lead, train, and coordinate the erection sequence. To streamline construction, the project was divided into three main zones (A, B, and C), aligning with the manufacturer's delivery sequence. Zones A and B overlapped, and installation began with these sections. However, when installation teams lack prior experience, project teams must account for learning curves that can impact efficiency. Panzica, the general contractor, encountered significant schedule delays due to subcontractors' inexperience with shaking out and setting materials. The situation was further complicated by harsh winter conditions, as the first delivery arrived in mid-February, bringing strong winds and heavy snow, further disrupting the erection sequence.

The table below presents a comparative analysis of productivity between mass timber and cast-in-place concrete, as observed on the INTRO project. Initially, the project was estimated and scheduled for construction using cast-in-place concrete, providing a solid baseline for comparison with the actual mass timber production. This comparison allows for an accurate evaluation of how the performance of mass timber stacks up against the original cast-in-place concrete plan. As shown, mass timber was erected in fewer working days (33 vs. 42), with significantly less manpower (daily average of 6.6 vs. 18 workers), while achieving higher productivity (2,322.5 vs. 1,824.9 square feet per day). These figures highlight the efficiency advantages that mass timber brought to the project, showcasing its potential for faster, more streamlined construction.

Mass Timber versus Concrete Productivity Rates

	Mass Timber	Concrete
Total zone floor area (SF)	76,644	76,644
Number of floors	6	6
Square feet per floor	12,774	12,774
Total working days	33	42
Total man-days	218	756
Total man-hours	1,744	6,048
Manpower average	6.6	18
Square feet per day	2,322.5	1,824.9
Square feet per man day	351.6	101.4

QUALITY

As one of the first major mass timber projects, maintaining a high focus on quality was of paramount importance to both the development and construction teams. However, being the first mass timber project that both teams had undertaken, the approach to ensuring the quality of the finished product was not entirely clear from the outset. One of the main quality considerations on this project was the potential of water damage on the exposed mass timber elements. Early on, there was significant debate regarding the use of manufactured-applied waterproofing techniques. These coatings would require a substantial upfront cost, quoted at over $400,000, to be applied by the manufacturer at the fabrication facility. Ultimately, the development team chose not to pursue this option and instead decided to implement on-site mitigation techniques.

This decision had a notable impact on the finished product. The team had to invest considerable time and resources repairing exposed beams, columns, and CLT panels that were stained due to water exposure from concrete pours, rain, and other construction-related water sources. While it's difficult to speculate on how effective the manufacturer-applied coatings would have been, it's reasonable to assume that the staining on the mass timber elements would have been significantly reduced had they been applied.

SUSTAINABILITY

One way of determining the sustainability of projects is the presence of third-party certification. Third-party certification represents the team's commitment to design and constructing a project that strives to meet sustainability thresholds. A clear correlation between mass timber structures and third-party rating systems have been drawn in the available data, as outlined in the sustainability chapter. This section will outline particular sustainability attributes of INTRO that were achieved and recognized within the LEED rating systems.

Like many mass timber projects outlined in the sustainability chapter, INTRO Cleveland's commitment to sustainability was codified in the form of LEED Gold status. Awarded on August 8, 2022, the project achieved this highly regarded level under the LEED BD+C: New Construction v4 rating system. The construction and development team achieved a respectable point total of 63 out 110 possible points within the rating system.

One of the key indicators of a project's sustainability is its third-party certification. This certification demonstrates the team's commitment to designing and constructing a project that meets rigorous sustainability standards. Available data, as discussed in the sustainability chapter, clearly show a strong correlation between mass timber structures and third-party rating systems. In this section, we will highlight specific sustainability achievements of INTRO Cleveland that were recognized within the LEED rating system.

Much like other mass timber projects featured in the sustainability chapter, INTRO Cleveland's commitment to sustainability was formally recognized with a LEED Gold certification. Awarded on August 8, 2022, the project achieved this prestigious status under the LEED BD+C: New Construction v4 rating system. The construction and development team earned an impressive 63 out of a possible 110 points within the rating system, underscoring their dedication to sustainable building practices.

The project emphasizes energy efficiency, carbon reduction, and environmentally responsible construction to meet high-performance building standards. Some of the sustainable choices made in this project include:

- Utilization of CLT and glulam beams from responsibly sourced wood, significantly reducing embodied carbon compared to traditional steel and concrete.
- High-performance building envelope with superior insulation to minimize heat loss.
- Efficient HVAC systems and LED lighting to lower operational energy use.
- Low-flow plumbing fixtures to reduce water consumption.
- Use of low-VOC (volatile organic compound) materials for healthier indoor air quality.

- Expansive natural lighting and ventilation to enhance occupant well-being.
- Located in a walkable, transit-accessible neighborhood, reducing reliance on cars.
- Integration of bike storage to support green transportation.
- Landscaping with native plants to minimize irrigation needs.

SAFETY

While there were no particularly noteworthy incidents regarding construction safety, this in itself is a significant achievement on a project of this scale. The limited number of accidents and injuries throughout the project is a testament to the team's strong commitment to safety management.

During the mass timber erection phase, potential safety concerns were proactively addressed through the strategic placement of numerous tie-off points for contractors. The team worked closely with Mass Timber Connections, a tie-off point and fastener manufacturer, to ensure that all tie-off points and hardware met fall protection requirements. Site inspections were conducted to verify that tie-off points were properly fastened with the correct hardware and that the appropriate number of tie-off points were in place. Furthermore, contractors were required to ensure they were securely tied off, or had fall protection in place, before installing any new points.

DISCUSSION QUESTIONS

Discussion Question 1

The Intro Cleveland project was groundbreaking for the mass timber industry in various ways. Outline some of the ways in which the project helped to break barriers within the industry and what it meant for subsequent projects.

Discussion Question 2

The decision to opt out of the manufacturer-applied water mitigation options affected the project in numerous ways. Outline all of the effects of that decision and discuss what you would have recommended for this project.

Discussion Question 3

In this case study, we highlighted the superior productivity of mass timber compared to cast-in-place concrete. What key factors contribute to the efficiency of mass timber construction, and what strategies can construction teams implement to further optimize productivity on mass timber projects?

REFERENCES

Binderholz. (n.d.). INTRO Cleveland, USA – Mass timber solutions. Retrieved March 23, 2025, from www.binderholz.com/en-us/mass-timber-solutions/intro-cleveland-usa/

CBCTC. (2020). Cleveland home to nation's largest mass timber project. The Cleveland Building and Construction Trades Council. www.cbctc.org/blog/cleveland-home-to-nations-largest-mass-timber-project

HP Architecture. (n.d.). Innovate Cleveland. Retrieved March 23, 2025, from https://hparchitecture.com/innovate-cleveland/

CASE STUDY II: ASCENT, MILWAUKEE, WI

Shortly after the completion of *INTRO Cleveland*, a groundbreaking mass timber project, another ambitious endeavor, *Ascent*, claimed the title of the tallest mass timber building in the world. Rising 25 stories, *Ascent* surpassed the 85.4-meter-tall *Mjöstarnet* in Norway by a full meter. Located in Milwaukee, Wisconsin, this mixed-use, multifamily building shattered previous expectations for the height potential of mass timber structures. However, the development team never set out to break the height record. Initially, building code restrictions limited them to just 19 stories. The team strategically leveraged local parking regulations, which permitted additional floors for parking. These floors were incorporated into the mass timber structure, allowing for an expansion in height. When they realized they were only a few meters shy of the world record, they made a slight adjustment to the roofline to claim the title.

This change was approved by the local municipality under the International Building Code's performance-based path for height approval, which offers flexibility beyond strict prescriptive requirements. To secure the approval, the team presented a compelling case, emphasizing the variability of mass timber regulations across different jurisdictions. This highlights the significant role that local building codes and municipal approaches can play in shaping the success of mass timber projects.

The *Ascent* project has earned several prestigious accolades, including the NCSEA Structural Engineering Excellence Award and the WoodWorks Wood Design Award for Innovation in 2023. These honors underscore both the innovative structural design of *Ascent* and the bold, risk-taking spirit of New Land Enterprises in pushing the boundaries of what mass timber can achieve.

PROJECT DETAILS

The 86.6-meter-tall tower consists of a structure made of CLT components, paired with glue-laminated timber beams and columns (Dreith, 2023). Like INTRO Cleveland, the podium consists of post-tension concrete, as well as the stair and elevator shafts.

Like the majority of large scale mass timber projects, the development, design, and construction teams were complex when compared to traditional structure. A wide array of professionals was needed to work through the design, procurement, and construction of this groundbreaking building. The main stakeholders responsible for successfully delivering this project include the following firms:

- **Developer:** New Land Enterprises
- **Contractor:** Swinerton
- **General Contractor:** C.D. Smith Construction
- **Structural Engineers**: Thornton Tomasetti
- **Architect**: Korb
- **Building Enveloper**: Rothoblaas
- **Timber Procurement:** Timberlab

The specifics of *Ascent's* structure are truly impressive. The building has a gross square footage of 456,000, with 324,000 square feet of that constructed from mass timber, perched atop a six-story concrete podium. *Ascent* is home to 259 residential units, complemented by ground-level retail space, and a wide array of amenities typical of a class-A multifamily building.

SCHEDULING

C.D. Smith, the general contractor for the *Ascent* project, has built a strong reputation for executing large-scale mass timber projects across the Midwest. Through their work on these projects, they've gathered valuable insights and data. According to C.D. Smith, if *Ascent* had been constructed with cast-in-place concrete, each floor would have taken eight to ten days to complete. In contrast, using mass timber, they averaged just five to six days per floor, with no need for curing or shoring. This allowed subsequent trades to begin work immediately.

More importantly, the team was able to install the windows and building enclosure right after erection, significantly reducing water exposure to the timber elements and enabling interior trades to work more efficiently.

Columns for each floor were flown to the level in a single pass, with 64 columns per floor, distributed across three picks. The lightweight mass timber columns allowed for up to 22 columns to be flown at once—something impossible with structural steel. Once the columns were placed on the floor, the installation team would proceed to set and fasten them in place. After the columns were in place, the beams were distributed evenly across the deck. Because the beams are larger than the columns, only one to two beams would be picked at a time, and they were set immediately adjacent to their installation locations. Each floor contained 75 beams, and the team averaged just one day to install all the beams per floor (Cover & Johansen, 2023).

In total, the construction of *Ascent* took about 22 months, compared to a projected 26 months if the building had been constructed using concrete.

ESTIMATING

New Land Enterprises, the development group behind *Ascent*, secured $130 million in financing from Bank OZK, equating to approximately $502,000 per rental unit. C.D. Smith, the general contractor, estimated that using mass timber added a slight premium of 1–2% to the overall budget. However, the rental premiums and shortened construction timeline more than offset this premium. Pre-leasing figures were impressive, with nearly 50% of the units rented before the building even opened.

For context, rental rates at *Ascent* start at $1,715 per month for one-bedroom apartments and $3,275 per month for two-bedroom apartments. The average rental rate is $34.98 per square foot, compared to $33.37 for the closest competitor (Gallun, 2023).

The team also provided a breakdown of the cost per square foot for just the timber elements of the project, which was $42 per square foot. For comparison, concrete—priced at a national low in the Milwaukee market—would cost $37 per square foot in Wisconsin (Mendez, 2022).

QUALITY

Quality concerns on *Ascent* were effectively managed and mitigated through an aggressive building enclosure schedule. By focusing on pre-planning, efficient erection processes, and strong subcontractor coordination, C.D. Smith ensured that the timber components were protected from the elements. The installation of exterior window assemblies immediately after the structure's erection led to earlier enclosure dates, significantly reducing the exposure of timber elements to weather.

In addition to closely sequencing the enclosure process, an extensive water mitigation plan was implemented to further protect the building.

C.D. Smith is a leader in the mass timber construction industry, thanks to the extensive number of projects they've completed. Their experience has been key in developing and applying comprehensive water mitigation techniques to prevent damage and boost efficiency. Experience, in fact, plays a critical role in determining the most effective quality control measures for a project.

SUSTAINABILITY

An interesting fact highlighted by the development team is that the lumber used for *Ascent* is replenished by natural growth in North American forests in less than 25 minutes, underscoring the renewable resource that wood provides for construction projects.

Beyond the impressive carbon emission savings, *Ascent* earned LEED Silver certification from the U.S. Green Building Council (USGBC) in 2022, adding a distinguished third-party rating to the project. The development team also quantified the carbon sequestration benefits of the timber frame, estimating the project's carbon emission savings at approximately 7,200 metric tons. This is equivalent to removing 2,100 cars from the road for a year or providing power to 1,000 homes for a year (Ascent, 2023).

In addition to the strong carbon data, the lightweight mass timber frame reduced construction waste and brought biophilic benefits, contributing to the building's overall environmental and aesthetic value.

DISCUSSION QUESTIONS

Discussion Question 1

Ascent was not originally designed to be the tallest mass timber structure ever built, yet it ultimately claimed that title. Discuss the project it surpassed, along with the key factors and challenges that contributed to Ascent achieving this record-breaking status.

Discussion Question 2

As highlighted in the scheduling section, the use of mass timber in this project led to significant time savings. What specific strategies can general contractors implement to further optimize the installation process and maximize productivity on mass timber projects?

Discussion Question 3

C.D. Smith's prior experience with mass timber construction played a crucial role in the successful delivery of this project. However, for a general contractor transitioning from traditional concrete or steel construction to mass timber, several challenges may arise. Discuss these challenges and explore strategies that can be adopted to mitigate them.

REFERENCES

Cover, J., & Johansen, C. (2023). World's tallest hybrid mass timber building construction: Ascent in Milwaukee. other. Retrieved 2025, from www.cdsmith.com/ascent#woodworks-video

Dreith, B. (2023, April 3). World's tallest timber tower replaced by forest growth "in less than 25 minutes." www.dezeen.com/2023/04/03/ascent-korb-associates-milwaukee-timber-revolution/

Gallun, A. (2023, October 30). In Milwaukee, a 25-story mass timber apartment building makes an ascent. https://urbanland.uli.org/sustainability/in-milwaukee-a-25-story-mass-timber-apartment-building-makes-an-ascent#:~:text=At%20$130%20million%E2%80%94or%20$502%2C000,white%20spruce%20imported%20from%20Austria.

Mendez, E. (2022, July 2). Ascent: A mass timber construction boom is coming to America. www.fastcompany.com/90756578/is-a-mass-timber-construction-boom-coming-to-america

Thornton Tomasetti. (2023). Ascent. www.thorntontomasetti.com/project/ascent

CASE STUDY III: T3 TOWER, MINNEAPOLIS, MN

Since its inception, mass timber products have enjoyed remarkable success in the European market. These materials made their debut in the United States in the early 2000s. However, it was the T3 tower in Minneapolis that marked a pivotal moment as the first commercial property in the United States to incorporate mass timber as both a structural element and a prominent feature in its interior design. The building showcases exposed glulam columns and beams, as well as nail-laminated timber (NLT) floors, which not only highlight the aesthetic appeal of the material but also demonstrate its potential for large-scale, sustainable construction.

PROJECT DETAILS

- **Type:** Commercial Office building
- **Developer:** Hines
- **Architect:** Michael Green Architecture (MGA), DLR Group
- **General Contractor:** Kraus-Anderson Construction Company
- **Structural Engineer:** Magnusson Klemencic Associates
- **Mass Timber Designer and Fabricator:** StructureCraft

Prior to 2016, there were no modern tall mass timber buildings in the United States. That changed with the completion of the T3 tower in Minneapolis, Minnesota. This seven-story, 220,000-square-foot high-rise, designed by MGA and DLR Group, was finished in 2016 and became the first tall mass timber structure in the United States as well as the largest mass timber building in the country at the time. T3, which stands for Timber, Technology, and Transit, represents a landmark project in the U.S. commercial building sector, demonstrating the viability of constructing tall buildings with mass timber.

T3 was built using glued-laminated timber (glulam) columns and beams, along with NLT floors, all sourced from trees killed by mountain pine beetles. The glulam columns and beams are made of spruce, while the NLT panels are crafted from SPF (spruce, pine, and fir). Most of the project's lumber came from trees in the Pacific Northwest region affected by the mountain pine beetle. Approximately 3,600 cubic meters of wood are used in the structure, including over 1100 8′ × 20′ NLT panels. T3's structural bays are 20 feet by 25 feet, with the timber beams spanning 25 feet and the 8-foot-wide NLT panels spanning 20 feet. This one-way system enabled building services to run parallel to the beams without clashes (Brownell, 2016).

To comply with building codes, the first floor of T3 was made of structural concrete. This was necessary to meet fire safety and other code requirements for taller wooden buildings. The concrete base provides additional fire resistance and structural support, allowing the upper floors to utilize mass timber. This lower concrete level houses retail spaces, a state-of-the-art fitness center, social workspaces, meeting areas, and a bike storage and repair station, complete with a complimentary community bike fleet. The building's exterior is clad in corrugated weathering steel, designed to reflect the character of the Minneapolis Warehouse District.

SCHEDULING

The structure was erected at a pace that far exceeds that of traditional steel or reinforced concrete buildings. The 221,552-square-foot mass timber building was completed in just 9.5 weeks. A total of 3,600 cubic meters of mass timber were assembled in this time, with each 30,000-square-foot floor completed in a remarkably efficient 9 days (MGA, 2020).

To support just-in-time (JIT) delivery to the jobsite, the mass timber supplier fabricated the NLT panels at a shop in Winnipeg, Manitoba, Canada, and then transported them to the jobsite, located 7 hours south of the plant, for delivery when needed. The decision to use NLT over other mass timber alternatives, such as CLT, was influenced by several factors, one of the most important being faster procurement time. Materials with a quicker procurement timeline are highly preferred because they reduce risks associated with delays in material delivery, disruptions to the project's critical path, increased costs due to inflation and other factors, as well as potential design changes that could introduce additional costs. JIT delivery is particularly effective with materials that can be procured quickly and reliably, ensuring that the project stays on schedule and within budget.

Glulam members on the T3 Minneapolis project were produced and supplied by a European partner (Hasslacher Norica Timber), an Austrian-based company known for its high-quality mass timber products. To streamline the JIT process, the European supplier first shipped the fabricated glulam members to the Winnipeg plant, where steel connection plates were installed. Once this step was completed, the components were then sent to the Minneapolis jobsite for installation as needed (StructureCraft, n.d.). The investment in JIT delivery was rewarded with the erection speed realized during construction.

COST AND THE ECONOMIC CASE FOR MASS TIMBER

Due to the novelty of the material, mass timber buildings are generally more expensive than traditional reinforced concrete or structural steel buildings. The total cost for the T3 Minneapolis project was estimated to be around $31 million (Alvarado & Henke, 2018). Investors have observed that mass timber buildings command some of the highest market rental rates. Although constructing with mass timber comes at a premium, this cost is often offset by the higher rental income these buildings generate. This is evident in testimony that performance buildings like T3 significantly outperformed pro forma estimates (Otter & Otter, 2023).

QUALITY

The designers of the T3 building estimate that its mass timber structure is 30% lighter than a comparable steel structure and 60% lighter than a comparable post-tensioned concrete structure. This reduction in weight enabled them to downsize the building's foundations, resulting in significant cost and schedule savings (Alvarado & Henke, 2018). T3's innovative design pushes the limits of modern architecture, seamlessly merging traditional craftsmanship with state-of-the-art technology. It highlights the potential of mass timber as a material for high-quality, high-performance buildings that meet the evolving demands of today's market.

T3 was also designed with a focus on energy efficiency, utilizing the thermal mass of wood to minimize energy consumption. Its sustainable features, along with the integration of advanced building systems, contribute to lower operational costs over the life of the project.

A truly quality product is one that goes beyond aesthetics to enhance the health and well-being of its occupants. The natural wood finishes in T3 create a warm, inviting atmosphere that is known to improve mood, reduce stress, and foster greater productivity. Exposure to natural materials, such as wood, has been linked to numerous psychological benefits, and T3's design incorporates these elements to promote a healthier, more vibrant work environment.

SUSTAINABILITY

The sustainability benefits of mass timber over conventional construction methods are thoroughly discussed in the sustainability chapter of this text. As previously mentioned, T3 utilized glulam columns and beams, along with NLT floors, most of which were fabricated from trees killed by mountain pine beetles, a fact that adds a unique element to the story. The project not only utilized trees that would have been lost anyway, it used trees sourced from responsibly managed forests, ensuring both sustainability and biodiversity preservation. The mass timber used in the building plays a crucial role in carbon sequestration, with the T3 building estimated to store around 3,200 tons of carbon throughout its lifecycle, as estimated by the design architect MGA (2020). This significantly offsets carbon emissions that would otherwise contribute to climate change.

It has been estimated that it would take just 15 minutes for all the forests within the United States and Canada to grow the amount of wood used to build T3. The environmental impact of choosing timber over traditional concrete is noteworthy: the project avoids 1,411 metric tons of CO_2 emissions, which would have been generated if the building had been made from concrete. This carbon offset is equivalent to removing 966 cars from the road for an entire year. Furthermore, the environmental impact of using wood in T3 is significant: the energy savings and reduction in greenhouse gas emissions from the timber used are roughly equivalent to powering a household for 430 years (Alvarado & Henke,

2018; Hickman, 2021). These figures underscore the substantial environmental advantages of incorporating mass timber into building design.

In addition to the building materials, T3 was thoughtfully designed with sustainability in mind, particularly in terms of transportation. Its location takes full advantage of transit-oriented and bicycle-first infrastructure, providing easy access to the Cedar Lake Bike Trail, commuter and light rail systems, and a well-connected pedestrian skyway system (Alvarado & Henke, 2018). These features encourage the use of alternative, eco-friendly transportation options, further aligning the building with its sustainability objectives.

T3 was awarded LEED Gold certification by the U.S. Green Building Council (USGBC) for achieving high standards in energy and environmental design. Some of the sustainable features that helped T3 achieve this certification include:

1. **Use of CLT:** T3 is one of the first large-scale buildings in the United States to use CLT as its primary construction material, which is a more sustainable and renewable alternative to steel and concrete.
2. **Energy Efficiency:** The building incorporates high-efficiency HVAC systems, LED lighting, and energy-efficient glazing that reduces overall energy consumption.
3. **Smart Building Technology:** T3 integrates advanced technology that monitors and adjusts the building's energy usage to optimize efficiency.
4. **Water Conservation:** The building is designed to reduce water consumption through low-flow fixtures and an efficient irrigation system.
5. **Transportation Accessibility:** It is in an area with access to public transportation, promoting sustainable commuting options for tenants.

Discussion Question 1

T3, the first tall mass timber building in the United States, was completed in 2016 in Minneapolis, Minnesota. Its construction is often credited with paving the way for the rise of tall mass timber buildings in the United States. Discuss the specific ways in which T3 helped spark interest in and advance the use of mass timber in the United States.

Discussion Question 2

T3 incorporated a range of design strategies that prioritized sustainability. Discuss the specific design considerations implemented in T3 that contributed to its overall environmental sustainability. How do these design choices align with broader sustainable building practices? Additionally, consider how the integration of these elements supports both environmental and social sustainability goals.

Discussion Question 3

Part of the mass timber used in the T3 project was sourced from young trees and trees killed by mountain pine beetles. Discuss the cost implications and sustainability benefits of using this type of timber for structural applications. How does the technology that enables the use of these trees contribute to both the economic feasibility and environmental advantages of mass timber construction? Additionally, consider the role of forest management practices in ensuring the responsible sourcing of such materials.

REFERENCES

Alvarado, I., & Henke, W. (2018). T3: Timber, transit & technology. Minneapolis, MN: University of Maryland. https://arch.umd.edu/programs/real-estate-development/colvin-institute-real-estate-development/competitions/capstone-competition/t3-timber-transit-technology-university-st-thomas

Brownell, B. (2016, November 8). T3 becomes the first modern tall wood building in the U.S. *Architect*. www.architectmagazine.com/technology/t3-becomes-the-first-modern-tall-wood-building-in-the-u-s_o

Hickman, M. (2021, October 19). U.S. Forest Service and Softwood Lumber Board launch a $2 million mass timber design competition. The Architect's Newspaper. www.archpaper.com/2021/10/us-forest-service-softwood-lumber-board-mass-timber-design-competition/

MGA. (2020, June 23). T3. MGA: https://mg-architecture.ca/project/t3-minneapolis/

Otter, C., & Otter, C. (2023, November 7). USA: Timber tops the market says T3 developer. WoodWorks. WoodWorks Engineered Wood Advisory Services. https://woodworks.events/usa-timber-tops-the-market-says-t3-developer/

StructureCraft. (n.d.). T3 Minneapolis. StructureCraft. https://structurecraft.com/projects/t3-minneapolis

CASE STUDY IV: 843 N SPRING STREET, LOS ANGELES, CA

Located in Los Angeles, California, *843 N Spring Street* stands as a testament to the versatility of mass timber products. The project showcases award-winning design paired with superior building performance. Technically a renovation of a 1980s warehouse in LA's historic Chinatown neighborhood, *Spring Street* blends cross-laminated timber (CLT) with steel bow trusses and concrete to harness the unique benefits of each material. The integration of green spaces was a key element in fostering the biophilic design features throughout the building.

Spanning 145,000 square feet, this mixed-use project offers a combination of retail, office, and amenity spaces for its occupants. Set on an existing foundation, the team utilized the substructure to accommodate parking needs for the facility. The design and functionality of the building earned the team numerous accolades, including:

- **Engineering News Record, California Best Projects**, Award of Merit—Office/Retail/Mixed-Use
- **Architect's Newspaper**, Best of Design Award—Honorable Mention, Commercial Office
- **Metropolis Magazine**, Planet Positive Awards—Best Design for Resilience
- **AIA Los Angeles**, Next LA Citation Award
- **LABJ Commercial Real Estate Awards**, Sustainable Project of the Year

These prestigious awards underscore the project's success in combining innovative design with sustainability.

PROJECT DETAILS

- **Building Type:** Commercial Office building
- **Client:** Redcar Properties Ltd.
- **Developer:** Redcar
- **Architect:** LEVER Architecture
- **General Contractor:** Shawmut Design and Construction
- **Structural Engineer:** Glotman Simpson US Inc.
- **Mass Timber Designer and Fabricator:**

Completed in 2024, the Spring Street project has unique design details that set it apart from other structures. The building rises five stories above a renovated concrete podium, using CLT and a hybrid steel frame for its structure. Floor heights range from 13 to 20 feet, contributing to the overall building height of 74 feet (Habchi and Smith, 2025). The lateral system between the two towers is composed of four bays of special concentrically braced frames (SCBF), transitioning to reinforced concrete shear walls (SRCWs). These design features were crucial in ensuring both the structural integrity and seismic performance of Spring Street. Above the concrete podium, level two is a Type 1-A suspended two-way flat concrete slab, providing fire separation and meeting code compliance requirements.

In line with sustainability goals, the project team made significant efforts to minimize the amount of demolition required for the existing structure. A key factor in reducing the need for additional foundations was the lightweight nature of the CLT panels. Without the reduced weight of the mass timber components, substantial modifications and additions to the existing concrete foundation would have been necessary. While additional foundations were still required to support the new structure, their scope was limited due to the reduced weight.

Mass timber also influenced the size and style of the structural members used in the building. The lighter weight of the material allowed the team to use round HSS braces for the SCBF frame. If heavier materials had been used, rectangular steel members and wide-flange braces would have been required, which would not have aligned with the design intent. This resulted in an 8% decrease in seismic mass compared to a traditional building, leading to an 11% reduction in the quantity of steel frames (Habchi and Smith, 2025).

SCHEDULING

The pre-construction and planning phase for 843 N Spring Street began in 2020 and extended into early 2021. This critical stage encompassed comprehensive site assessments, feasibility studies, design development, and structural evaluations of the existing warehouse foundations. The team also navigated the complex permitting and entitlement process, ensuring alignment with local regulations and project goals.

Demolition and site preparation commenced in mid-2021 and continued into late 2021. Selective demolition was carefully carried out to preserve and reinforce the existing foundations, minimizing environmental impact and retaining the site's historical integrity. Concurrently, site grading, utility installation, and preparations for the mass timber construction were completed, setting the stage for the next phase of development.

Between late 2021 and early 2022, the foundation work and structural modifications took place. The team reinforced and adapted the existing foundations, installed structural steel, and constructed the concrete core. Coordination efforts were made for the delivery and assembly of the mass timber components, marking a significant milestone in the project's structural evolution.

From mid- to late 2022, the erection of mass timber elements began. This phase included the installation of CLT panels and glulam beams, along with the integration of structural and mechanical systems. Additionally, the building's façade and glazing elements were added, completing the exterior shell of the building.

The interior build-out occurred between late 2022 and early 2023, focusing on the installation of mechanical, electrical, and plumbing (MEP) systems, interior partitions, flooring, and finishes. Sustainability remained a central priority throughout this stage, with the incorporation of energy-efficient systems and low-carbon materials that helped reduce the environmental impact of the construction process. The project was completed in 2024, following final testing of building systems, obtaining necessary city approvals, and facilitating tenant move-ins and operational handovers.

QUALITY

The quality of 843 N Spring Street is evident in its thoughtful design, sustainable practices, premium materials, and meticulous attention to detail throughout construction. These elements came together to create a building that not only met but exceeded expectations for environmental responsibility, aesthetic appeal, and functionality.

The design balanced aesthetic elegance with functional performance, prominently featuring mass timber, including CLT and glulam beams. This choice provided structural integrity while fostering a warm, inviting interior. During demolition, selective techniques preserved and reinforced the existing foundations, maintaining key structural elements. This innovative approach strengthened the building and reduced its carbon footprint, contributing to its overall sustainability.

Energy efficiency was prioritized throughout the project. The building was outfitted with state-of-the-art HVAC systems, energy-efficient LED lighting, and other green technologies to minimize energy consumption. Mechanical, electrical, and plumbing (MEP) systems were designed for operational efficiency and long-term reliability. Interior finishes, including flooring and partitions, were selected for both durability and aesthetic appeal, enhancing the building's overall quality. Additionally, locally sourced and sustainable materials were chosen to further elevate the building's environmental and structural integrity.

Upon completion, the project underwent comprehensive testing to ensure all systems were fully operational. Rigorous quality control measures confirmed that the mechanical, electrical, and plumbing systems performed as intended, and the building met all necessary standards for occupancy, safety, and sustainability.

SUSTAINABILITY

Although the Spring Street project did not pursue third-party certification to officially recognize its commitment to sustainable building practices, it seamlessly integrated a broad range of green construction techniques and gathered vital data regarding the building's sustainability performance. Notably, Spring Street is an adaptive reuse project, where the design team repurposed an existing building for a new purpose. This approach not only preserved the site's historical and cultural significance but also played a key role in minimizing waste and reducing energy consumption. The project specifically maintained the original masonry and concrete structure of a former two-story commercial building. By retaining the existing foundation and structural elements, the team significantly reduced CO_2 emissions and enhanced construction efficiencies (Los Angeles Business Journal, 2024; Shawmut Design and Construction, 2024). While much of the original structure was preserved, some additional foundational work was necessary. The team reinforced the existing concrete superstructure by doweling into it, ensuring structural integrity and supporting the new design elements.

843 N Spring Street was awarded Gold in the 2024 Commercial Real Estate Awards, and recognized as the top project in the sustainability category by the Los Angeles Business Journal. The project was praised for its integration of biophilic design and low-carbon construction practices, notably the use of CLT panels, which sequestered an estimated 1,357 metric tons of carbon (Los Angeles Business Journal, 2024). This innovative approach highlights 843 N Spring Street's commitment to environmental responsibility and its contribution to sustainable architecture in Los Angeles.

DISCUSSION QUESTIONS

Discussion Question 1

The Spring Street project is unique in a multitude of ways. One of which was the limited disturbance and high utilization rate of the existing foundations, from a previous warehouse structure completed in the 80s. Outline what design and materiality factors played a role in the limited need to demolish and/or alter the existing foundations.

Discussion Question 2

How does the incorporation of cross-laminated timber (CLT) in the Spring Street project enhance both the structural efficiency and sustainability of the building? What specific challenges and trade-offs might arise from using mass timber in a hybrid design that combines CLT with steel and concrete elements, particularly in terms of seismic performance, material selection, and long-term durability?

Discussion Question 3

Although the project did not receive third-party sustainability certification, what sustainability initiatives did the design and construction team implement to enhance the building's environmental performance? Discuss the strategies employed to maximize its environmental benefits.

REFERENCES

Habchi, R., & Smith, C. (2025). Mass Timber for 843 N. Spring Street. Structure. www.structuremag.org/article/mass-timber-for-843-n-spring-street/

Los Angeles Business Journal. (2024). 2024 Commercial Real Estate Awards: Sustainability. Los Angeles Business Journal. https://labusinessjournal.com/business-journal-events/2024-commercial-real-estate-awards-sustainability/

Shawmut Design and Construction. (2024, March 7). 843 N Spring Street wins sustainability project of the year from *Los Angeles Business Journal.* www.shawmut.com/news/843-n-spring-street-wins-sustainability-project-of-the-year-from-los-angeles-business-journal

CASE STUDY V: HIGH LINE TIMBER BRIDGE, NEW YORK CITY, NY

Located in the heart of New York City, the High Line—Moynihan Connector Bridge was designed to address a critical urban infrastructure need. Spanning 600 feet, this elevated pedestrian pathway creates a seamless link between the iconic High Line park and the newly renovated Moynihan Train Hall. The project improves pedestrian access and connectivity, establishing a vital connection between Manhattan's West Side neighborhoods and key transportation hubs. By enhancing mobility and facilitating access to public spaces, the Moynihan Connector plays a key role in shaping the city's dynamic urban landscape.

PROJECT DETAILS

- **Type:** Elevated Pedestrian Bridge
- **Developer:** Empire State Development, Brookfield Properties, and Friends of the High Line
- **Architect:** Skidmore, Owings & Merrill (SOM) in collaboration with James Corner Field Operations
- **General Contractor:** Turner Construction
- **Structural Engineer:** Thornton Tomasetti and Aspect Engineers
- **Mass Timber Designer and Fabricator:** Mercer Mass Timber

The High Line—Moynihan Connector consists of two distinct bridges designed to enhance both connectivity and the urban experience:

- **Woodland Bridge**: A 340-foot-long path running east along 30th Street, featuring deep soil beds that support a lush and vibrant landscape. The design includes 63 trees, 90 shrubs, and over 5,200 grasses and perennials. The bridge's green corridor creates a natural shield for pedestrians, offering a serene escape from the surrounding urban environment and a sanctuary amidst the city's hustle and bustle.
- **Timber Bridge**: A 260-foot-long warren truss structure, constructed using sustainably sourced Alaskan yellow cedar glulam beams. This bridge spans Dyer Avenue, connecting the Woodland Bridge to Magnolia Court at Manhattan West. The use of glulam timber reinforces the project's commitment to sustainability while providing a robust and visually striking structure.

The project was made possible through a public–private partnership involving Empire State Development, Brookfield Properties, and Friends of the High Line. It resulted from a collaborative design effort between Skidmore, Owings & Merrill (SOM) and James Corner Field Operations, blending architectural innovation with thoughtful landscape design (Empire State Development, 2021). With an estimated budget of $50 million, construction began in early 2022, and the bridge officially opened to the public in June 2023 (Empire State Development, 2021).

Beyond its functional role as a pedestrian route, the Moynihan Connector Bridge enriches the urban fabric of New York City. By integrating sustainable materials, green spaces, and innovative design, the project offers a peaceful retreat while enhancing connectivity on Manhattan's West Side.

SCHEDULING

The High Line—Moynihan Connector project, which began its design phase in January 2021, was initiated by New York State with the goal of creating a continuous pedestrian pathway linking the High Line, Manhattan West, and Moynihan Train Hall (Empire State Development, 2021). By the end of 2021, the design phase concluded, with Skidmore, Owings & Merrill (SOM) and James Corner Field Operations finalizing the plans (SOM, 2023). In September 2021, the project secured funding, and preparations for construction began, culminating in a groundbreaking ceremony on February 24, 2022, marking the official start of construction (Empire State Development, 2021).

Construction activities commenced in March 2022, focusing on foundation work, including excavation and piling to support the bridge structures. By May 2023, the Timber Bridge, a 260-foot-long Warren truss made from sustainably sourced wood, was installed over Dyer Avenue. The bridge's glue-laminated wood beams were assembled at street level and then craned into place, with the full bridge assembly weighing 256,704 pounds (The High Line, 2023). On June 22, 2023, the Connector was officially opened to the public, providing a vital pedestrian connection across the urban landscape (The High Line, 2023).

The project's two distinct bridge structures, the Woodland Bridge and the Timber Bridge, each featured unique construction techniques. The Woodland Bridge, a 340-foot-long bridge running along 30th Street, is supported by deep soil beds that sustain a lush landscape of trees and plants, with weathering steel columns and angled bracket arms dynamically adjusting to varying soil depths (SOM, 2023). In contrast, the Timber Bridge, a 260-foot-long structure, uses a Warren truss design made from sustainably sourced glued laminated timbers. This design required minimal ground connections, preserving the surrounding infrastructure (SOM, 2023).

The Woodland Bridge's superstructure comprises precast concrete panels supported by a weathering steel structure, creating a continuous planter that supports large trees typical of an Eastern Deciduous Forest (SOM, 2023). The Timber Bridge, on the other hand, consists of a three-span truss made of glued laminated timbers, each measuring approximately 8 feet tall and 3 feet wide (Thornton Tomasetti, n.d.). Landscaping for the Woodland Bridge included deep soil beds for 63 trees, 90 shrubs, and over 5,200 grasses and perennials, enhancing the green corridor (The High Line, 2023), while the Timber Bridge's use of timber complements the natural elements of the High Line and preserves the urban environment beneath (SOM, 2023).

COST AND FINANCING

The High Line—Moynihan Connector project had an estimated budget of $50 million, funded through a strategic public–private partnership. The financial backing came from three major contributors: Empire State Development (ESD) provided a substantial $20 million grant, accounting for 40% of the total funding; Brookfield Properties matched this contribution with another $20 million, or 40%; and Friends of the High Line contributed $10 million, representing the remaining 20% (Empire State Development, 2021). This diverse funding structure was key to making the ambitious project a reality.

The financial allocations were carefully distributed to cover various aspects of the project, ensuring that both immediate construction and long-term sustainability were addressed. Direct construction costs amounted to $30,848,457, while indirect construction costs, which included elements such as site preparation and utilities, totaled $10,443,154. Soft costs, including expenses related to environmental reviews, design, legal fees, and marketing services, were estimated at $7,789,696. Additionally, the project accounted for $918,693 in maintenance and operational costs, reflecting the ongoing care necessary to keep the connector in good condition post-construction (Empire State Development, 2021).

These funds were essential in supporting the construction and design of critical infrastructure components, including the Woodland and Timber Bridges, as well as extensive landscaping and improvements to existing infrastructure. By carefully allocating resources to these key areas, the project team ensured the creation of a sustainable, functional, and aesthetically impactful addition to the High Line that would serve the public for years to come.

QUALITY

The project team took a multifaceted approach to ensure the highest quality in the design and construction of the High Line Timber Bridge. Central to the success of the project was the careful selection of materials, which directly influenced the bridge's strength, durability, and overall longevity. The 260-foot-long timber bridge was built using a warren truss design, a structure renowned for its robust strength and long-lasting performance, ensuring that the bridge would withstand the test of time. The selection of glulam timber played a critical role in ensuring the bridge's stability and longevity. With superior strength-to-weight ratios compared to traditional timber, glulam allowed for a more efficient, durable, and sustainable structure. This material was particularly effective in maintaining the bridge's structural integrity without compromising on the natural aesthetic (Thornton Tomasetti, n.d.).

A key consideration in the design was minimizing the impact on the surrounding environment. By reducing the number of ground connections, the team preserved existing roadways and infrastructure, maintaining the integrity of the surrounding area. This minimal ground impact approach also made it possible to keep disruption to a minimum during construction, reflecting a conscientious effort to balance modern engineering with environmental respect (SOM, 2023). In terms of assembly and installation, the team demonstrated extraordinary precision and expertise. The glulam truss of the Timber Bridge was carefully assembled at street level before being lifted into place over Dyer Avenue, a feat that showcased the meticulous engineering required to bring the project to life (Thornton Tomasetti, n.d.).

As previously stated, the project also featured the 340-foot-long Woodland Bridge, designed to accommodate varying soil depths along the pathway. This was achieved through deep soil beds reinforced by weathering steel columns and dynamically angled bracket arms, which provided the necessary stability while adjusting to the shifting contours of the landscape. This thoughtful consideration of the site's specific needs contributed to a safer, more secure structure. Precast concrete panels further enhanced the bridge's structural integrity. These panels, integral to the construction, provided additional durability and strength, ensuring that the bridge could withstand the wear and tear of time and traffic (The High Line, 2023). The use of weathering steel in the structural components was another smart design choice. Known for its resistance to corrosion, weathering steel would significantly reduce the bridge's long-term maintenance needs, offering a solution that combined strength with efficiency (SOM, 2023).

The inclusion of natural elements further enhanced the safety and comfort of the bridge. As mentioned before, the Woodland Bridge was designed to feature deep soil beds that supported a variety of trees and plants, which not only contributed to the aesthetic appeal but also created a natural barrier that shielded pedestrians from the harsh winds and noise of the surrounding urban environment. This thoughtful integration of nature into the design made the bridge not only safer but more inviting, ensuring that those who crossed it could enjoy a peaceful, secure, and immersive experience in the heart of the city (The High Line, 2023).

In addition to its robust design, the bridge also prioritized accessibility and safety. The Connector is fully ADA-compliant, ensuring that individuals with disabilities can navigate the pathway with ease. By elevating the pathway above the busy streets, the bridge provides pedestrians with a safer, more scenic route that minimizes interaction with vehicular traffic, fostering a sense of security as people traverse the space (Empire State Development, 2021).

SUSTAINABILITY

The High Line Timber Bridge was conceived as a natural extension of the original New York City High Line, a visionary adaptive reuse project that reimagined an old railway system into a vibrant elevated park. The High Line's inception was rooted in sustainability, seeking to transform an underused urban infrastructure into a green space that promotes pedestrian access, encourages wildlife, and fosters healthier living in the city. The adaptive reuse of the site was intended to provide a sustainable alternative to urban sprawl while contributing to the city's environmental goals. With the addition of the Timber Bridge, the project continued to build on this ethos, extending the High Line's mission to create green, accessible spaces that align with sustainability principles.

A key element of the High Line's design has always been its integration of nature into the urban landscape, and this continued with the Timber Bridge. The bridge was carefully crafted to incorporate natural elements in a way that aligned with the original goals of the High Line. Deep soil beds along the path support an impressive range of flora, including sixty-three trees, ninety shrubs, and over 5,200 perennials and various grasses. These plantings not only provide habitats for wildlife but also work to mitigate the heat island effect in the city, reducing ambient temperatures and improving air quality. This natural vegetation helps cool the surrounding urban environment, offering a reprieve from the concrete sprawl below (The High Line, 2023).

The timber bridge itself exemplifies the project's commitment to sustainability. The bridge features a truss system crafted from Alaskan yellow cedar, a material chosen not only for its aesthetic beauty but also for its environmental benefits. Alaskan yellow cedar is a sustainable resource that contributes to the reduction of greenhouse gas emissions, as it sequesters carbon dioxide during its growth cycle, helping to offset the carbon footprint associated with construction (Simpson, 2024). This choice aligns with the broader sustainability goals of the High Line, ensuring that the bridge contributes positively to the environment rather than detracting from it.

Beyond the selection of materials, the design of the bridge also addresses stormwater management in a sustainable and innovative way. The bridge features a permeable walkway design, with open joints that allow rainwater to flow freely through the surface and cascade down into the soil beds below. This design not only aids in reducing the strain on the city's stormwater infrastructure but also provides the plants with a natural source of irrigation. By allowing rainwater to nourish the plants directly, the bridge eliminates the need for supplemental watering systems, further conserving water resources and reducing the environmental impact of the project.

In this way, the High Line Timber Bridge serves as both a functional and symbolic extension of the original High Line vision. It continues the legacy of adaptive reuse, utilizing natural materials and design principles that support a sustainable urban environment. By incorporating elements like mass timber, permeable surfaces, and integrated green spaces, the bridge enhances the overall ecological and social benefits of the High Line, further solidifying it as a model of urban sustainability.

DISCUSSION QUESTIONS

Discussion Question 1

Mass timber is typically used in building construction, but its distinctive properties also make it an attractive option for bridge construction. Discuss the characteristics of mass timber that make it particularly well-suited for bridge design and construction, with a focus on pedestrian bridges, such as the one discussed in this case.

Discussion Question 2

How does the addition of the High Line Timber Bridge align with the original sustainability goals of the High Line project, and in what ways does it contribute to urban ecology, climate resilience, and the promotion of pedestrian-friendly spaces? Consider the integration of natural materials, the use of mass timber, and the permeable design in your response.

Discussion Question 3

How do the design and material choices for the High Line Timber Bridge, such as the use of Glulam timber, weathering steel, and minimal ground impact, reflect the importance of balancing structural integrity, environmental sustainability, and construction efficiency in modern urban projects?

REFERENCES

Empire State Development. (2021). High line moynihan connector civic project. https://esd.ny.gov/high-line-connector

Simpson, L. B. (2024, June 26). High line – moynihan connector. SOM. www.som.com/projects/high-line-moynihan-connector/

SOM. (2023). High line – moynihan connector. www.som.com/projects/high-line-moynihan-connector/

The High Line. (2023). High line – moynihan connector. www.thehighline.org/connections/

Thornton Tomasetti. (n.d.). High line moynihan connector. Retrieved March 14, 2025, from www.thorntontomasetti.com/project/high-line-moynihan-connector

CASE STUDY VI: WALMART'S HEADQUARTERS

Considered a landmark project for both the State of Arkansas and the mass timber industry, the completion of Walmart's world headquarters in Bentonville marks several significant "firsts" for the sector. The project, slated for phased openings from 2020 through 2024, features more than 2.5 million square feet of mass timber construction. It includes over eleven individual buildings, such as offices, a fitness center, multiple dining areas, a daycare facility, an auditorium, and a visitor center. Spanning more than 350 acres, the campus will integrate walking and biking trails. With over 1.7 million cubic feet of regionally sourced lumber, it is the largest mass timber campus project in the United States (Gensler, 2024).

This groundbreaking project—North America's largest mass timber campus—was partially made possible by Walmart's investment in the manufacturing company responsible for producing the material used in the construction.

PROJECT DETAILS

The construction attributes of Walmart's new home offices are impressive. Completed in 2024 the campus encompasses over 350 acres of developed land, and boasts over 3,000,000 square feet of new buildings (King, 2024). Utilizing CLT and glue-laminated timber the type III-B project is impressive in scale.

- **Design Architect, Office Buildings and Executive Architect, Overall Campus Design:** Gensler
- **Architect, Walton Family Whole Health & Fitness:** Dudal Paine Architect
- **Architect, Little Squiggles Children's Enrichment Center:** Page
- **Landscape Architect:** SWA

- **Urban Design:** Sasaki
- **Civil, Traffic:** Walter P. Moore
- **MEP, Sustainability:** Arup
- **Structural:** Fast & Epp
- **Lighting:** ISP Design
- **Associate Architect, Sam Walton Hall and 8th & Plate Food Hall:** Miller, Boskus & Lack
- **Architect, Coffee Shop and Water Features**: Marlon Blackwell Architects
- **Design Architect, Hotel and Welcome Center:** 5G Studio Collaborative
- **Architect of Record, Hotel and Welcome Center:** Boka Powell
- **MEP, Hotel and Welcome Center:** HP Engineering
- **Architect, Central Utility Plants:** Nielsen Architecture
- **Program Manager:** Turner Construction
- **Parking:** Walker Consultants
- **Civil:** CEI Engineering Consultants
- **Mobility:** Nelson\Nygaard and Velocity NWA
- **Trails:** Trailblazers
- **Sustainability:** William McDonough + Partners
- **Accessibility:** American Construction Investigations
- **BIM:** Willow
- **Acoustics**: SLR
- **Accessibility:** ACI
- **Food Service:** Ricca Design Studio

SCHEDULING

A project of this scale requires significant time for development and construction. Given its size, the project team opted for a phased approach, with distinct sections and buildings opening for occupancy throughout the year. The vision for this cutting-edge campus was conceived in 2019 (Joyner, 2019), and Walmart has reported that over 10 million project hours have been managed to date (Ross, 2025).

Construction officially began in 2019 with demolition, infrastructure preparation, site grading, and utility installation. The first building completed was the new Layout Center, which was finished in the fourth quarter of 2020. The phased approach allowed Walmart employees to occupy the Layout Center early, while construction continued on other amenity structures. To ensure the successful integration of new buildings with planned phases, work on parking, paving, and power infrastructure provided the necessary resources for employees.

By 2022, the focus shifted to developing innovative amenities, including the vertical construction of the Walton Family Whole Health and Fitness Center, Sam Walton Hall, and multiple parking decks. Design and material procurement also progressed throughout the fiscal year. By 2023, vertical construction accelerated, and the efforts of the design and construction teams began to take shape, highlighted by the official opening of the fitness center and other amenity spaces. In 2024, the continued opening of office spaces and the completion of final parking facilities signaled that the campus project was nearing its final stages.

ESTIMATING

An interesting aspect of this project is Walmart's strategic investment in Structurlam, the manufacturing company responsible for producing the mass timber. This cost was an important factor in the overall project expenses. The Canadian-based firm, Structurlam, was selected for its high-quality timber production, and Walmart's investment was driven by the goal of supporting local production. Initially, the team considered sourcing raw materials from out of state, but this option became less attractive both from a cost and sustainability perspective. Ultimately, Walmart invested $90 million in a state-of-the-art manufacturing facility near the project site, enabling Structurlam to open a facility in Arkansas that sourced locally grown wood from surrounding areas.

The team purchased a 288,000-square-foot former steel manufacturing facility, which was retrofitted to produce mass timber products in 2021 (Dalheim, 2019). Structurlam's first customer was Walmart, which received over 1.1 million cubic feet of cross-laminated timber (CLT). This investment not only made financial sense for the project but also created over 100 jobs and bolstered the local economy. While the total project cost remains proprietary, the investment highlights Walmart's long-term commitment to sustainability and local economic development.

Walmart's economic policy of Every Day Low Cost (EDLC) was evident throughout the development of the project. Emphasis on phased construction, integrated into the annual budgeting process, ensured a manageable cost allocation and reinforced the company's focus on efficient, sustainable development.

QUALITY

Walmart's acquisition of Structurlam provided benefits beyond just monetary gains. It played a crucial role in strengthening the team's relationship concerning procurement, manufacturing, and quality control expectations. Established in 1962, Structurlam has built a strong reputation for high-quality production, a reputation that was reflected throughout the project. By acquiring the manufacturer, Walmart gained greater control over the quality process, ensuring consistency and precision at every stage.

SUSTAINABILITY

The project team integrated several sustainability initiatives into the design of the 350-acre campus, with walkable and bikeable paths as a primary consideration for site layout. Ensuring connectivity between structures while encouraging users to engage with nature was central to the design process. The campus was planned with a neighborhood approach, featuring four distinct quads. The design evolved organically through land typology, internal roads, and landscaping. A key element was maintaining appropriate building heights to ensure a "human-centric" scale across the campus (Eberhardt, 2025). This theme was consistently applied across all buildings to create continuity and prevent an overwhelming presence.

The human experience was at the core of the project. To support health and well-being, the design incorporated extensive walkable and bikeable paths, alongside a state-of-the-art fitness facility for employees. The landscaping includes 750,000 newly planted native flowers and approximately 5,000 trees.

Walmart's commitment to sustainability was recognized with LEED Platinum certification, the highest level a project can achieve. This prestigious designation highlights the development team's innovation in sustainable design. Additionally, the use of mass timber contributed to a 15% reduction in carbon emissions compared to traditional building materials (Downes, 2025).

The transportation infrastructure included over 300 EV charging stations and 1,000 bike parking spots. A rentable bike fleet provides employees and visitors with a convenient way to navigate the seven miles of walking and cycling paths. Autonomous lawn care vehicles are also employed to maintain the property.

In addition to site features, several sustainable building strategies were implemented. Notably, dynamic glass products were selected to actively reduce energy consumption in the buildings.

DISCUSSION QUESTIONS

Discussion Question 1

Walmart pushed the boundaries in a number of ways, including but not limited to unique financial partnerships and cutting edge sustainability initiatives. Outline and discuss these initiatives that have been mentioned throughout the case study.

Discussion Question 2

How did this project affect the local and state economies of Arkansas? Present the impacts a project like this can have on local economies.

Discussion Question 3

What allowed Wal-Mart to have direct control over the quality control process in regard to manufacturing? Discuss what important element of this project allowed this to happen.

REFERENCES

Dalheim, R. (2019, December 10). Walmart invests in mass timber maker Structurlam, paves way for $90 Million plant. www.woodworkingnetwork.com/news/woodworking-industry-news/walmart-invests-mass-timber-maker-structurlam-paves-way-90-million#:~:text=Walmart%20invests%20in%20mass%20timber,$90%20million%20plant%20%7C%20Woodworking%20Network

Downes, S. (2025, January 22). A look inside Walmart's sustainable new US headquarters. *Sustainability Magazine*. https://sustainabilitymag.com/articles/a-look-inside-walmarts-sustainable-new-us-headquarters

Eberhardt, E. (2025, January 24). Gensler uses mass timber at an "unprecedented scale" for Walmart HQ. www.dezeen.com/2025/01/21/walmart-mass-timber-arkansas-hq-gensler-swa/

Gensler. (2024). Walmart home office: Projects. www.gensler.com/projects/walmart-home-office

Joyner, S. (2019, August 29). Gensler to design Walmart's new 350-acre campus in Arkansas. https://archinect.com/news/article/150155863/gensler-to-design-walmart-s-new-350-acre-campus-in-arkansas

King, D. (2024, December 12). Walmart home office – mass timber win project. www.woodworksinnovationnetwork.org/projects/walmart-home-office

Ross, J. (2025, January 21). Walmart's new home – massive timber project shatters records. https://woodcentral.com.au/walmarts-new-home-massive-timber-project-shatters-records/

Index

www.ingramcontent.com/pod-product-compliance
Ingram Content Group UK Ltd.
Pitfield, Milton Keynes, MK11 3LW, UK
UKHW020422070226
467784UK00013B/217
9781032769141